FREGE IN PERSPECTIVE

FREGE

IN PERSPECTIVE

JOAN WEINER

CORNELL UNIVERSITY PRESS

Ithaca and London

First published 1990 by Cornell University Press.

International Standard Book Number 0-8014-2115-2
Library of Congress Catalog Card Number 89-28377

Printed in the United States of America

*Librarians: Library of Congress cataloging information
appears on the last page of the book.*

⊛ The paper used in this publication meets the minimum
requirements of the American National Standard for Permanence of Paper
for Printed Library Materials Z39.48—1984.

To
Bess-Paula Weiner
and
Pearl Raskin Miller

Contents

vii

Acknowledgments

Frege's writings are largely responsible for my commitment to philosophy. When I entered Harvard's graduate program in philosophy in 1975, I did so with divided loyalties. Mathematics, the subject of my undergraduate major, and mathematical logic, in particular, still had a hold on me. I began graduate school planning to become a philosopher who did her research in mathematical logic. But my education did not proceed as planned. Harvard did not place the obstacles in my way, I did. I felt that I needed to commit myself to one field and let go of the other.

On Burton Dreben's advice, I approached Frege's work with a question that, he suggested, might help me make such a commitment. Why did Frege, a mathematician, need to write something as philosophical as *Foundations of Arithmetic*? It was my pursuit of this question that transformed me into a philosopher and that has guided much of my subsequent work. Along the way, I have received a great deal of invaluable help.

The first help was from Dreben, who not only suggested the question but undertook to challenge me on my answers. Goaded by him week to week, I discovered what he was trying to help me find—not an interpretation of Frege, but how to speak in my own philosophical voice.

My early work on Frege was also marked by the influence of Warren Goldfarb's lectures on Frege in his 1977 course on early analytic philosophy. Remnants of the sensibility Goldfarb imparted survive even now. I owe to those lectures my strategy of beginning this study of Frege with a careful reading of the first chapter of the *Foundations of Arithmetic*. And I owe to them some of the details of that reading.

The members of my dissertation committee, which included Hilary Putnam as well as Dreben and Goldfarb, made further contributions to the development of my approach to Frege's writings through their repeated challenges to my interpretations and arguments. I cannot adequately acknowledge the help I received from each of them. Naturally, none of them and, indeed, none of those whose help I acknowledge below, is to be held responsible for the doctrines the book espouses—with one possible exception. I would like to explain.

Perhaps the most controversial part of my interpretation of Frege's writings has to do with the way the views expressed in "On Concept and Object" influence my interpretation of Frege's other works. I argue that the doctrines of Frege, the champion of precision, commit him to a view on which the ineffable plays a central role. But for a long time I could not decide whether Frege would have acknowledged this consequence of his doctrines or whether I thought such a consequence could possibly be acceptable.

The former issue, I am convinced, cannot be settled. The latter, however, was settled almost accidentally. In 1983, I had the great good fortune to come to know, and to begin studying with, a consummate musician with a talent for pedagogy so prodigious as to be almost indistinguishable from magic. From Raquel Adonaylo I was able to learn the importance of important nonsense.

My work on this book was supported, during the summers of 1985 and 1986, by National Science Foundation grant #SES 8509959 and, during the 1985–86 academic year, by a Mellon Fellowship at the University of Pittsburgh. My year at Pittsburgh provided me with an ideal environment in which to write. I am grateful to the many faculty members and graduate students of the philosophy department there who made me so welcome during that year. In addition, I owe thanks for useful discussions to Nuel Belnap, Robert Brandom, Kenneth Manders, Alexander Nehamas, and Nicholas Rescher.

Since that time, I have acquired many other debts. In my revisions I was able to make use of detailed and helpful comments from Cora Diamond, Michael Resnik, and an anonymous referee. Among the others who have influenced my work on this book, through either conversations or correspondence, are Anil Gupta, Paul Horwich, Philip Kitcher, Hans Sluga, and Crispin Wright.

I am especially grateful to John Ackerman of Cornell University Press for the indispensable help, encouragement, and unfailing courtesy he has furnished me.

Other debts have accrued over periods of longer duration that do not fit so neatly into the present chronology. The first I want to mention is the debt I owe to the constant support of my colleagues in the philosophy department at the University of Wisconsin–Milwaukee. They saw to it that I was provided with course reductions and leaves when I most needed them and that the income from the Mellon Fellowship was supplemented so that I would not have to make any financial sacrifice. A happier, better-run department is unlikely to be found.

I also thank Thomas Ricketts, a philosophical influence and friend of long duration. I have benefited greatly from discussion and correspondence with him and from his extensive comments on the penultimate draft of this book.

My final, and greatest, thanks go to Mark Kaplan, who read and commented in detail on every draft of every chapter, whose criticism was always honest and always available—often on a moment's notice—and who never lost faith in the project.

JOAN WEINER

Milwaukee, Wisconsin

A Note on Presuppositions, Conventions, and Translations

I N recent years the literature on Frege has become increasingly technical, some of it readable only by specialists immersed in disputes
concerning the extensive secondary sources, the technical details of his
notation and constructions, or the appropriate translation of his writings. Although I am not writing an introduction to Frege's writings, this
book is also not addressed solely to specialists. I shall therefore take the
opportunity here not only to introduce some of the abbreviations and
conventions I use, but also to indicate the background I am presupposing.

Long before I had any acquaintance with Frege's writings, outside of
reading "On Sense and Meaning" (or "On Sense and Reference," as it
was known then) in an introductory philosophy of language course; I
had been taught that Frege introduced the first version of the logical
notation used today, that he attempted to show that arithmetic was
analytic by proving the truths of arithmetic from the truths of logic, and
that this proof failed because one of his logical laws, Basic Law V, was
inconsistent with the others. More generally I thought of Frege as
having introduced contemporary philosophical ways of thinking about
language and mathematics as well as logic. I have tried to write a book
that presupposes no information about Frege's writings other than this.
As I mentioned above, however, this is not an introduction to Frege's
works. I have attempted neither to catalog and explain his major views
nor to introduce and evaluate generally accepted interpretations.

In the interest of making this book accessible to the nonspecialist, I have confined most of the discussion of secondary sources, as well as the more intricate textual interpretations and most uses of logical notation, to footnotes and appendixes. Also, since virtually all Frege's writings are available in translation, my citations are generally to the translations only, and I have for the most part quoted only the English translations. In a few cases I have found it necessary to refer to untranslated passages from *Basic Laws* and hence the only citation is to a German text. There are, however, a few exceptions that must be explained here.

The issue of how Frege's term *Bedeutung* and its cognates should be translated into English has become a matter of controversy. Although 'meaning' is used in most translations currently in print, until recently the more popular choices were 'reference' and 'denotation'. Since part of my aim is to explain how Frege's use of this term is to be understood and since each of the translations mentioned here can appear to presuppose a particular explanation, I have often chosen to leave *Bedeutung* untranslated and to treat it as an English term. I apologize for the infelicity of expression which results. In some of the passages I have quoted there seemed to me to be a potential for confusion, and I have indicated, in these cases, when an English word is meant as a translation of *Bedeutung* or one of its cognates. Here and in other quoted translations when I thought it would be helpful to include a German phrase or word, I have cited both the English translation and a German publication.

Another of Frege's terms which has been left untranslated and treated as an English term is 'Begriffsschrift'. It has no natural English translation, and the convention of using the German has already been established in Montgomery Furth's translation of *Basic Laws*. Following Furth, I have used the term unitalicized to talk about Frege's logical notation and, in some cases, about his logical system. When it is italicized I am employing it as an abbreviation of the title of the work in which Frege's logic was first introduced, *Begriffsschrift, eine der arithmetischen nachgebildete Formelsprache des reinen Denkens*.

Frege's logical notation presents another problem. Although it is, in a very important sense, the ancestor of the logical notations employed today, the actual symbols Frege used are quite foreign to contemporary philosophers who are not Frege scholars. I have not tried to introduce and explain his symbols, since I have very little need for them here. My concern is not with the formal details of Frege's accomplishments but with the underlying philosophical view. In a few discussions, when it is necessary to talk about Frege's logical regimentation, I have made use of contemporary logical symbols. Only in a very few technical analyses that appear in footnotes and in one appendix have I talked about and used Frege's actual symbols. These discussions are addressed to specialists and can be skipped by most readers.

Abbreviations

Frege's works are cited in text and notes by these abbreviations plus the page numbers.

BLA   *The Basic Laws of Arithmetic*, trans. Montgomery Furth

BS   *Begriffsschrift und andere Aufsätze*. Ed. Ignacio Angelelli

BEG   *Begriffsschrift, a Formula Language, Modeled upon That of Arithmetic, for Pure Thought*. Trans. Stefan Bauer-Mengelberg

CN   *Conceptual Notation and Related Articles*. Trans. Terrell Ward Bynum

COR   *Philosophical and Mathematical Correspondence*. Ed. Gottfried Gabriel et al. Trans. Hans Kaal

CP   *Collected Papers on Mathematics, Logic, and Philosophy*. Ed. Brian McGuinness. Trans. Max Black et al.

FA   *The Foundations of Arithmetic*. Trans J. L. Austin

GGA   *Grundgesetze der Arithmetik*

KS   *Kleine Schriften*. Ed. Ignacio Angelelli

NS   *Nachgelassene Schriften*. Ed. Hans Hermes, Friedrich Kambartel, and Friedrich Kaulbach

PW   *Posthumous Writings*. Ed. Hans Hermes, Friedrich Kambartel, and Friedrich Kaulbach. Trans. Peter Long and Roger White

TWF   *Translations from the Philosophical Writings of Gottlob Frege*. Ed. Peter Geach and Max Black

WB   *Wissenschaftlicher Briefwechsel*. Ed. Gottfried Gabriel et al.

FREGE IN PERSPECTIVE

Introduction

BEGINNING in 1879, with the publication of his *Begriffsschrift, eine der arithmetischen nachgebildete Formelsprache des reinen Denkens* (*Begriffsschrift, a Formula Language, Modeled upon That of Arithmetic, for Pure Thought*, hereafter *Begriffsschrift*), Gottlob Frege's work, published and unpublished, was almost entirely devoted to one project. That project was to define the individual numbers and the concept of number and to show that all the truths of arithmetic can be proved using only these definitions and the general laws of logic. Although they contain some discursive writing, two of his three major publications, *Begriffsschrift* and *Grundgesetze der Arithmetik* (*Basic Laws of Arithmetic*, hereafter *Basic Laws*), are primarily books of proofs. In *Begriffsschrift* Frege set out the laws of logic, some derivations, and some logical definitions that would later play a role in his definitions of the numbers. In the first volume of *Basic Laws* (1893) he introduced a slightly modified version of his logic, and in the second volume (1903) he attempted to show that, using this logic, all the truths of arithmetic can be proved from definitions alone. His other major publication, *Die Grundlagen der Arithmetik: Eine logisch-mathematische Untersuchung über den Begriff der Zahl* (*The Foundations of Arithmetic: A Logico-Mathematical Enquiry into the Concept of Number*, hereafter *Foundations*) (1884), is a philosophical discussion of his project. That project, of course, was a failure. In order to prove the truths of arithmetic in *Basic Laws*, it was necessary to add a law of logic which had not originally appeared in *Begriffsschrift*. But the result of adding this new law was to make Frege's logical system inconsistent.

Although the project ultimately failed, much contemporary philosophy has been influenced by Frege's writings. In fact, this influence has been so profound that it is difficult to believe there can be any mystery about Frege's views. Frege's slogans and the terms he coined have become an integral part of the thought of contemporary analytic philosophy. I am convinced, however, that this situation has led many philosophers to ignore, or to mistake the importance of, some of Frege's truly mysterious claims. The result is a distorted picture of the general outlines of his thought.

The problem, I think, is that we customarily view Frege's work from too close a range. This closeness is constituted by the assumption that Frege is truly one of us. My aim is to tell a different story about Frege's overall picture by looking at Frege from a greater distance, by stepping back and entertaining the possibility that he is not one of us. It is this distancing—this stepping back from an object whose character is misapprehended if viewed from too close a range—that I mean to convey by the title *Frege in Perspective*.

There are difficulties inherent in such an approach. Even if one does not assume that contemporary concerns are obviously the most interesting, the advantage of attributing contemporary concerns to Frege is that his work requires no further motivation: if Frege's concerns are our concerns, then surely we understand the motivation and significance of his work as well as we understand the motivation and significance of our own. If Frege's concerns are not ours, however, and if one wants to claim that Frege was a deep and serious thinker, more is required. A story must be told on which these concerns seem compelling. But where can such a story be found? There are two possible sources of information: the actual historical context in which Frege wrote (the work of his contemporaries or precursors), and the text he produced.

It may initially seem that the more promising strategy is to start with an investigation of the historical philosophical or mathematical context of Frege's work. For Frege's most sustained discussion of the motivation for his project is found in the early sections of *Foundations*, the passages that fit most easily with the view of Frege as a twentieth-century analytic philosopher. Thus, to gain distance from his work, it may help to view his project as addressing concerns of nineteenth-century German philosophers—concerns that will seem foreign to us. This strategy seems promising for other reasons as well. For it would be absurd to suppose that Frege's views developed entirely uninfluenced by those of his contemporaries. And, although Frege does not discuss his motivation in great detail, some of his contemporaries are more forthcoming about what motivates their own projects. If their discussions can serve to explain what motivates Frege's project as well, these discussions may enable us to come up with a satisfying story about why Frege wrote what he did.

In *Gottlob Frege*, Hans Sluga undertakes to exploit this possible source of insight into Frege's thought.[1] Sketching an account of the philosophical climate of nineteenth-century Germany, Sluga discusses the influence of Leibniz and Kant on Frege's contemporaries as well as the views against which Frege's contemporaries were reacting. He also identifies some of the philosophical works Frege read and some striking similarities between statements in those works and Frege's statements of his own views. Sluga's work makes it quite clear that Frege's questions did not arise in a vacuum. It also gives rise to an interesting interpretation of Frege's corpus—an interpretation with which I have a great deal of sympathy. Nonetheless, I remain unconvinced by Sluga's claims about how Frege's contemporaries influenced him. The problem is that the evidence needed to establish these claims is simply unavailable.

The small size of his surviving philosophical and mathematical correspondence as well as the scarcity, in Frege's published and unpublished writings, of discussions of the works of others creates obvious difficulties for drawing conclusions about the extent to which Frege's contemporaries influenced him. But the way Frege discusses the works of others creates a far worse difficulty. These discussions, whether they appear in published or unpublished writings, and whether his attitude is approving or disapproving, are invariably superficial. In addition, the writers whose views Frege discusses most seem to be those with whom he has a fundamental lack of sympathy. Frege devotes considerably more pages to the work of Hilbert, for instance, than to the work of almost any other philosopher or mathematician. Although Frege's discussions of Hilbert are very revealing if our interest is in Frege's own views, they can tell us little of interest about the relation of Frege's and Hilbert's views. Frege exhibits a total lack of regard for, or understanding of, what Hilbert was trying to do.

In fact, Frege is notorious not only for the lack of charity with which he assessed the work of his contemporaries, but also for his conviction that his own work was entirely misunderstood. The evidence provided by Frege's remarks about his contemporaries and his discussions of

[1]Hans Sluga, *Gottlob Frege*.

their work suggests that Frege—a deep and subtle thinker who, although he took some classes in philosophy as a student, received his training in mathematics—had his own idiosyncratic understanding of the philosophical terms and slogans of his time. It suggests that his concerns, and his understanding of the philosophy of his time, were rather different from those of his contemporaries—even from those of his teachers. But if what the evidence suggests is in fact true, then the similarity in manner between the ways Frege and his contemporaries express certain of their respective views is *just* a similarity in manner: Frege was a philosophical maverick who assigned his own meaning and significance to philosophical claims that were in the air. And if this is so, then no account of the philosophical and/or mathematical climate of his time can help but fall short of providing us with a comprehensive picture of the concerns Frege meant his project to address.

But all this is only what the evidence *suggests*. In fact, because of the lack of conclusive evidence, I would not be willing to argue that Sluga's attributions of philosophical lineage are wrong. It seems consistent with the available evidence that Sluga is right. But it also seems consistent with that evidence that he is wrong. The assumption that Frege's project *must* have been related in a significant way to the projects of his contemporaries is simply that—an assumption. And it is an assumption that can just as easily blind us to important threads in Frege's thought as the assumption that Frege's concerns are our concerns. My strategy is to avoid assuming the correctness of either of these characterizations of Frege's background. Instead, I mean to see what conclusions can be drawn from concentrating on Frege's texts alone.

The burden imposed by this strategy is to make the familiar look unfamiliar. For the view of Frege as one of us has been derived from what seems a straightforward and obvious reading of his text. If this same text is to reveal a Frege with foreign views, our approach to it must differ from the usual approach. And so my approach does.

I begin by focussing on the motivation for Frege's overall project. His first contribution to this project is *Begriffsschrift*, in which he intended to set out the laws of logic and to give some of the definitions that were to be used in his later attempt to define the numbers. Although the overall project is mentioned in the preface to *Begriffsschrift*, that preface affords little discussion of the project's motivation. The pretensions of his project are given their first sustained discussion in the early sections of *Foundations*, the work that appeared between the publication of *Begriffsschrift* and that of the *Basic Laws*. I begin my investigation with a detailed reading of these sections.

At first blush, they do not look to contain anything novel or surprising. To those of us who come to his writings with preconceptions about our Fregean heritage, the motivation offered seems transparent, natural, and in need of no amplification. My technique for making this motivation seem less natural is to concentrate on the oddness of Frege's use of certain notions that look, on the surface, to be either everyday or familiar philosophical notions. My aim is to show that Frege's understanding of these notions is actually extremely quirky and that the success of his motivation is dependent on the quirks. So viewed, there is more to these sections, and to Frege's motivation, than a cursory glance suggests.

The early sections of *Foundations* contain a series of arguments, some apparently trivial, others apparently aimed at refuting substantive views held by other philosophers. Although it is common to examine these arguments individually, I read them as contributions to a unified argument. All the component arguments, on my reading, are parts of a larger argument that the notion of number is not primitive. It is, on my reading, crucial to the success of Frege's motivation that he be able to give a convincing argument that this notion is not primitive. But what is required of an argument that some notion is not primitive? This question cannot be answered without an interpretation of Frege's understanding of primitiveness.

The purpose of my reading of chapter 1 of *Foundations* as consisting of a unified argument is to highlight the central roles played by several notions, among them primitiveness and definability, in Frege's motivation. Frege does not indicate that there is anything unusual or controversial about his understanding of these notions. Nor does he suggest that he is introducing anything new. I will argue, however, that attention to Frege's use of these notions exhibits gaps and tensions in his motivation.

Thus, the aim of my detailed, almost line-by-line, reading of these sections is not to replace a natural and familiar motivation with one of greater philosophical depth. Rather, the aim is to exhibit the gaps and tensions to which I've just alluded. To make these sections work—to derive from them a credible motivation of Frege's overall project—it is necessary to address these tensions. The depth that I ultimately find in Frege's work I find not in the early sections of *Foundations*, but in some of his later works when read with these tensions in mind. In fact, my reinterpretation of the early sections of *Foundations* is, in many respects, a flatter, more modest reading than most readers are inclined to give to these sections of *Foundations*.

Most of chapter 1 of *Foundations* is taken up with discussions of Kant and Mill. After introducing the question "What is the number one?" Frege attempts to refute Kant's and Mill's answers. It is common to take these arguments seriously, to try to determine whether or not Frege has succeeded in his attempts to refute Kant and Mill. Although I have devoted a great deal of space to these discussions of Kant and Mill, I have not attempted to adjudicate between Frege's views and those of Kant and Mill. In fact, not only have I ignored Kant's and Mill's actual views, I have not even tried to present rounded pictures of Kant-as-Frege-read-him or Mill-as-Frege-read him. Frege's opponents emerge, on my reading, as cardboard figures.

Part of my reason for refusing to read substantive and plausible views into Kant (Mill)-as-Frege-read-him is my conviction that this would result in an inaccurate reading of the content of Frege's actual remarks. The sheer brevity of Frege's writings speaks against one who would like to find in Frege well-rounded readings of other philosophers. Consider Frege's treatment of Kant. Frege not only claims to acknowledge great debts to Kant; he also uses citations from Kant as part of the motivation of his project. Yet Kant is discussed on only sixteen pages of *Foundations* and two pages of *Begriffsschrift*, and not explicitly anywhere in the early papers on *Begriffsschrift*. Such discussions can hardly provide evidence on which to attribute to Frege a rounded picture of Kant, one of the most subtle and complex figures in the history of philosophy.

Still, in spite of my reluctance to attribute a substantive reading of Kant to Frege, I have chosen to include a rather lengthy discussion of Frege's views on Kant. My motive is quite simple: I am convinced that the philosophical motivation for Frege's overall project appears in his comments about Kant. And I am convinced that this motivation underlies much of the detail of Frege's philosophical picture. The influence of Kant on Frege's picture, however, can only be the influence of Kant as Frege interprets him—the view of Kant expressed in the total of less than eighteen pages of discussion. Thus, my discussion is not of Kant's views but of Kant's words as appropriated and interpreted by Frege. Following Frege, I quote Kant rarely and out of context. That is, I have limited my discussion of Kant to the quotations that Frege cites placed in the context in which Frege placed them.

In spite of this lack of evidence in Frege's text, my reading may still seem to commit me to attributing a substantive interpretation of Kant to Frege. For many of the passages I want to take seriously are passages in which Frege avows an allegiance to Kant. One might argue that, while not all readers need attribute an interpretation of Kant to Frege, any

reader who takes these passages seriously must do so. My own interpretation of Frege's early writings, which seems to attribute to Frege
nothing more than a commitment to certain apparently Kantian slogans, may seem uncharitable—especially since Frege himself was such
a deep thinker. But not all deep thinkers are deep readers. Frege's
claims need not have been based on a substantive grasp of Kantian
philosophical theory, and there is no textual evidence that Frege was a
deep reader of anyone else's work.

It seems even clearer that there is no rounded picture of Mill-as-
Frege-read-him. Outside of *Foundations*, there are only four brief mentions of Mill in Frege's published work and one mention in his unpublished work. Frege's actual discussions of Mill are scattered
throughout the first thirty-eight pages of *Foundations*. The most sustained of these discussions occupies less than five pages. Most of
Frege's comments about Mill are sarcastic and ungenerous. Frege's Mill
has all the earmarks of a straw man invented for purely polemical
reasons.

On my view this is easily explained. As I read chapter 1 of *Foundations*, it constitutes Frege's argument that the notion of number cannot
be primitive. Frege's strategy is to win our conviction by providing a
survey, and detailing the defects, of all the possible ways one might be
tempted to characterize number as a primitive notion. For polemical
purposes, he has an interest in being able to identify those who have
been tempted. Lifting seemingly appropriate passages from *System of
Logic*, he attributes some of these characterizations to Mill. But Frege
makes no serious attempt to understand Mill's view.

What makes Frege's discussion of Mill important is not what it tells us
about how Frege understood Mill, but what it tells us about how Frege
understood primitiveness and definability. Given how little Frege says
about how he understands these notions, one might be tempted to
think that there is no hidden depth to Frege's use of them. But by my
lights, a detailed reading of Frege's discussion of Mill dispels any such
impression. For that reading reveals that the effect of his discussion is to
place a set of constraints on these notions. These constraints do not
apply to any everyday or traditional understanding of primitiveness or
definability. And the odder Frege's understanding of these notions
looks, the odder his project of defining the numbers begins to look.

Part 1 of this book consists of three chapters. The first, as I have noted
above, consists of a careful reading of the introduction and first chapter
of *Foundations*. I try to show that, once one takes seriously the con-

straints Frege's criticisms of Mill place on the notion of primitiveness, along with the constraints other discussions in the early sections of *Foundations* place on some of Frege's other notions, the flat reading of Frege's motivation begins to exhibit gaps and tensions not originally apparent. In the second and third chapter I investigate the extent to which some of the explicit passages in Frege's texts succeed in filling some of these gaps. For this purpose, I concentrate on Frege's writings from 1884 (the period of *Foundations*) and earlier. On certain issues, for instance that of Frege's understanding of definition, I rely on later texts as well. But I try to argue, when I do this, that the later views are consistent elaborations of the views in his earlier writings. The verdict is that, while most of the obvious gaps can be filled in by attention to the explicit words on Frege's pages, tensions remain.

This verdict has some interesting consequences, which I explore in parts 2 and 3. In part 2, I begin to examine the extent to which chapter 2 of *Foundations*, as well as some of Frege's later writings, provides responses to some of these tensions. In part 3, I discuss the significance of attributing these responses to Frege. I argue that the introduction of these responses allows Frege a characterization of his project which fills in the gaps in the motivation provided in *Foundations*. Although Frege does not recharacterize the overall project in his published work, he does this in some of his later unpublished writings. The most explicit of these is a remarkable paper, "Logic in Mathematics," which seems to include discussions of all the major themes of his work. The characterization of Frege's project which appears in "Logic in Mathematics" is entirely in accord with both his early writings, as I read them, and the demands introduced by reading his later writings as responses to tensions in the early writings. Although the view I attribute to Frege on the basis of the early sections of *Foundations* is a shallow one, the philosophical picture that emerges from reading some of his later writings as answers to tensions in that view is both deep and compelling. This is one of the features of my interpretation which, I think, helps to make its revisionism plausible. Another is that the interpretation provides an explanation of certain traditional puzzles concerning the interpretation of Frege's writings. For instance, the so-called concept *horse* problem with which so many of Frege's readers have struggled is not, on my reading, a problem for Frege.

At this point, I may seem to have outlined a rather odd interpretive strategy. I have said that the interpretation for which I argue in parts 2 and 3 results from reading some of Frege's later works as responses to tensions in some of his earlier works. But most of these tensions are never explicitly addressed in his writings. Since there is little evidence that Frege was aware of these tensions I might well have ended my discussion of Frege's work with a series of critical notes. However, I think it is interesting and important to examine how Frege might have responded to some of the difficulties caused by these tensions. And I think it is especially interesting that, although Frege never discusses many of these tensions, his later works suggest responses to them.

But are these responses Frege's? Would Frege have agreed with the responses I outline for him? Without explicit textual evidence that Frege worried about the problems I discuss and addressed himself to these problems via the later works I discuss, it would be illegitimate for me to make claims about how Frege, the historical person, would have felt about these responses. And, in most cases, such evidence is not available. Thus, in some of the discussions in part 2 and in most of the final chapter, I have explicitly shifted my emphasis from the views of Frege-the-historical-person to the views to which Frege's writings commit him.

This is not to say that parts 2 and 3 concern the views of a mythical philosopher of my own invention. I do mean to argue that Frege-the-historical-person's views commit him to these responses by philosophical reasoning that must be acceptable if everything else Frege said was acceptable. And I do think that Frege would have accepted this reasoning. What I do not know is whether Frege would have accepted the conclusions drawn. Frege might well have chosen to revise some of the views with which he began rather than accept the conclusions to which they commit him.

To give up a view in the face of unacceptable consequences was, after all, not foreign to Frege. Frege's initial assumption was that the truths of arithmetic were licensed by no more than what licensed logic (by what he later called 'the logical source of knowledge', *die logische Erkenntnisquelle*). Thus, he attempted to define the numbers and prove the truths of arithmetic by means of these definitions and logical laws alone. Yet he came to realize that the proofs he required depended on a law, Basic Law V, which introduced an inconsistency into his Begriffsschrift. Frege's response was to retreat from his initial view that arithmetic could be known through the logical source of knowledge. The inconsistency of Begriffsschrift with Basic Law V is, of course, a matter of deductive consequence. But his conclusion that arithmetic cannot be known through the logical source of knowledge was not. He could have

drawn other conclusions. Whether, faced with some of the consequences of his views which I discuss, Frege would have opted for
revision or affirmation simply cannot be known.

It should thus be clear that my decision to go beyond what I am
confident Frege-the-historical-person would have avowed and to examine views to which Frege had, probably unbeknownst to himself, committed himself, is not a decision to instruct Frege on what he should
have said but did not. Nor is it a decision to loosen the rules of interpretation, to let the imagination run wild and speculation run free. My
object in writing this book was to provide a reading that accounts for the
words on Frege's pages. I would regard a convincing argument that
there are passages that contradict either views I attribute to Frege or
views to which I claim Frege is committed as damaging to my interpretation. Thus, my distinction between the views of the historical
person and the views to which the historical person committed himself
is one born of caution. It is a product of my unwillingness flatly to
attribute to Frege doctrines that I am not confident he would acknowledge as his own. It is nonetheless my opinion that some of the views I
am reluctant to attribute to Frege-the-historical-person (the views on
elucidation discussed in chapter 6, for instance) are the only views that
can make sense of some of Frege's writings.

But do they really make sense? In contrast to the Frege who has left us
with a legacy of philosophical terms and issues, Frege, as I interpret
him, uses a vocabulary of words that (to us) have odd meanings derived
from a foreign and dated philosophical framework. And the picture I
attribute to Frege is not just different from most of those offered by
Frege scholars. Frege's picture, on my reading, turns out also to be
deeply flawed. How, then, can I claim to have made better sense of the
words Frege wrote than someone who reads Frege as having introduced a productive twentieth-century philosophical agenda?

If the legitimacy of an interpretation of the work of a philosopher of
historical prominence depends on its attributing to that philosopher a
philosophical project in which philosophers are engaged today, then
my claim to have made sense of Frege's writings is surely illegitimate.
But no historian could seriously propose such a general condition of
adequacy on interpretation. What would it do to our understanding of
Descartes, to our understanding of Plato? If it is tempting to apply the
criterion in the case of Frege it is only because we are so used to viewing
him as a contemporary.

But the fact is that no claim to lineage can be known a priori. It must

be recognized as possible that Frege does not bear the relation to contemporary analytic philosophy he is said to bear—that the influence of Frege is really the influence of Frege-misread. That is, it must be recognized as possible that some reading of Frege's work that denies our projects Frege's historical cachet—that attributes to him concerns with which we lack sympathy—is more accurate to Frege's texts.

Thus, to argue, as I have done, that Frege is to be interpreted in a way that makes him a far less familiar philosopher than we are wont to think him is an entirely legitimate attempt to deny certain claims of intellectual influence. But it is not just that. My interest in historical scholarship does not primarily stem from an interest in claims about intellectual lineage. It is mostly a product of my conviction that there is at least as much of genuine interest in unfamiliar modes of thought as in familiar ones. In rendering Frege unfamiliar—even in rendering him guilty of fatal errors (do we really think our successors will look back at us as divinely excused from such indignities?)—I do not think I have rendered him any less profound, any less interesting. On the contrary.

Frege, as I read him, does not simply look to be an implausible candidate for admission into the ranks of contemporary analytic philosophy. He turns out to be positively hostile to some of the most prominent views attributed to him, and widely held, by our philosophical peers. On my interpretation, the criticism of empiricism which emerges from Frege's writings is a criticism that does not simply threaten the view at which Frege levels his attack, but also presents a rather fundamental challenge to contemporary empiricists of the Benacerraf or Kitcher mold.[2] On my interpretation, Frege emerges as a writer for whom contemporary versions of platonism, the doctrine with which he is perhaps most often—and frequently with approval—associated, turn out to be virtually empty views. Far from providing a cachet to the contemporary framework in which the philosophy of mathematics is conducted, my Frege raises serious questions about the motivation and substance of the assumptions that underlie that framework.

But I do not mean to suggest that my interest in Frege is solely or even primarily in the extent to which he shows that contemporary analytic philosophers get things right or wrong. The interest of a historical figure cannot be determined simply by assessing the use of her or his writings in solving contemporary problems. If Frege, as I read him, has

[2]See, for instance, Paul Benacerraf, "Mathematical Truth"; Philip Kitcher, *The Nature of Mathematical Knowledge*.

anything to contribute to us today—and I think he does—it is not by way of introducing philosophical problems for us to solve, solutions to contemporary philosophical problems, or even methods by which philosophical problems can be solved. What Frege has to offer us is a model of philosophical virtue. Almost everyone who has grappled with Frege's writings has been moved by Frege's intellectual honesty. But Frege's philosophical virtue goes far beyond his courageous ability to understand the extent of the devastation wreaked by Russell's counterexample to Basic Law V. His virtue is also exhibited in the way his explicit views fit together into a philosophical picture of extraordinary depth and subtlety. And it is exhibited in the creativity of his responses, both explicit and implicit, to tensions in his views.

As should be clear by now, I expect that my interpretation of Frege will be controversial. Thus, it may seem only appropriate, given the many Frege scholars who have long struggled with these texts, that I should defend my interpretation with a detailed critical discussion of some of the interpretations available in print. I have, however, chosen to avoid extensive discussion of the secondary literature. I have done this mostly in the interest of clarity and accessibility. The outlines of Frege's project, as I read it, are complex. To add a detailed discussion of the rich and diverse literature on his work would, I am convinced, make this book inaccessible to all but specialists. Thus, in the interest of not obscuring the story I want to tell and of making this story accessible to the general philosophical public, I have omitted extensive exegesis and criticism of most secondary sources. But while my interpretation may be controversial, I do not want to suggest that I owe no debts to the secondary literature. I have tried to make some of these debts clear in footnotes.

It may seem that I have not addressed the real cost involved in providing a revisionist interpretation of Frege's work. As long as the concerns, terminology, and issues of contemporary analytic philosophy can be regarded as Frege's legacy to us, as long as Frege can be regarded as having staked out positions in contemporary debates, his writings can be of considerable use to us in our attempts to formulate and defend our own positions. His writings have indeed been put to this use. Citations of his work appear routinely in papers whose subject matter concerns not the interpretation of his texts, but philosophical worries about language, mathematics, or logic. To someone who believes that Frege has outlined, and contributed to our understanding of, projects on which we are working, it may seem that a revisionist reading of

Frege cannot help but make Frege's work less important. It may even seem that if Frege-reinterpreted makes no contribution to the philosophical endeavors to which Frege is generally thought to have contributed mightily, then the project of reinterpreting Frege is an exercise in obstructing philosophical progress.

But the cost is not as high as this suggests. A reinterpretation of Frege's work, no matter how compelling, need not prevent philosophers from learning from Frege-misread if Frege-misread helps in either articulating or solving philosophical puzzles. Nor need a convincing reinterpretation cast doubt on the motivation of contemporary philosophical investigation. Appeals to authority cannot suffice to motivate a serious philosophical worry. If a contemporary worry is of genuine interest, the recognition that Frege did not share it should cast no doubt on the legitimacy of pursuing that worry. The only effect the existence of a convincing reinterpretation of Frege's work need have on those who view themselves as his heirs is the relinquishment of this particular view of their heritage. It is thus difficult to see why any revisionist reading of Frege, no matter how successful, should have the effect of obstructing philosophical progress.

I offer these comments in part as a response to Michael Dummett's vehement arguments to the effect that a revisionist approach to Frege's writings is illegitimate, and even intellectually vicious.[3] It is not a detailed response. Dummett's contention depends on the recognition of a distinction between textual exegesis and philosophy. And while I do not recognize the existence of such a distinction, I see no point in engaging here in an abstract discussion of whether or not such a distinction can be drawn. Nor do I think there is much to be gained from an abstract discussion of the legitimacy of reinterpreting Frege. Ultimately, a revisionist approach to Frege's writings can be vindicated only by particular revisionist interpretations that make contributions to philosophical thought. Thus, the way to make a case for the legitimacy of *my* revisionist interpretation is to make a case for the accuracy of my reading and for the importance of the philosophical achievement of Frege, as I read him. It is to the making of that case that I now turn.

[3]These arguments appear in the chapter titled "Principles of Frege Exegesis" in Michael Dummett, *The Interpretation of Frege's Philosophy.* They also appear in Dummett's "An Unsuccessful Dig," a review of G. P. Baker and P. M. S. Hacker, *Frege: Logical Excavations.*

PART I

CHAPTER 1

"What Is the Number One?"

T HIS book is a sustained attempt to trace one path through Frege's writings from his *Begriffsschrift* on. It is not a discussion of a series of related issues that appear in Frege's work. Nor is the path it traces historical. Its starting point is not the chronological starting point. My conviction is that virtually all of Frege's writings can be viewed as contributions to a unified project, and my interpretation is an attempt to explain this project and the contributions made by the particular pieces of writing. I have begun by addressing the questions that such an interpretation of Frege's work must answer. What was Frege's overall project? What was Frege's motivation for undertaking this project? And I have addressed these questions by examining the earliest writings in which Frege himself attempts a lengthy answer—the introduction and first chapter of *Foundations*.

Frege begins *Foundations* by asking what the number one is. He devotes the first two sections of chapter 1 to arguments for the mathematical importance of his project of undertaking to answer this question, and he devotes the second two sections to arguments for its philosophical importance. The remainder of chapter 1 consists of criticisms of the views of others. On a cursory reading, the discussions in the motivational sections, as well as Frege's criticisms of others, seem both transparent and natural, especially to a twentieth-century analytic philosopher—whose philosophical ears have been trained with the phrases that appear there. Thus, my problem, in trying to give a reading of these sections, is similar to Frege's problem in trying to explain what the number one is. Both efforts are aimed at audiences to whom the answer seems too obvious or familiar. Frege says, "The first prerequisite for learning anything is thus utterly lacking—I mean, the knowledge that we do not know" (FA iii). Frege's strategy for dealing with his expositional problem is to begin by arguing that we really do not know what the number one is. My strategy for dealing with my expositional problem will be to begin by arguing that we really do not understand Frege's explicit motivation. I will not, however, begin with an examination of the discussions of Frege's motivational sections in the secondary literature. Instead, I will begin with discussion of the texts themselves. For if Frege is truly one of the clearest of all philosophical writers, then perhaps we ought to be able to take Frege at his word in these sections.[1] I will argue that if we do this, then, in order to provide a convincing motivation for his project, it will be necessary to address some serious questions that Frege does not address in these sections.

Frege initially attempts to motivate his project by appealing to mathematical worries. It seems that, if we are to take Frege at his word, we should take seriously not only Frege's claims about the motivation of his project but also the fact that Frege begins with a mathematical rather than a philosophical story about the importance of his project.[2] Furthermore, it does not seem unreasonable to view his project as a mathematical project, for it looks similar, in many respects, to other exclusively mathematical projects. I will argue, however, that Frege's project is actually very different from most other, apparently similar, projects undertaken by mathematicians. Indeed, the mathematical motivations often attributed to Frege do not really provide motivations for his actual

[1]This seems to be Michael Dummett's view. He suggests, in the preface to *The Interpretation of Frege's Philosophy*, that the divergence of opinion concerning what Frege was up to, among people who write on Frege, is unnecessary because Frege's writings are so clear. After calling Frege "one of the clearest of all philosophical writers," Dummett says, "Frege's standard of lucidity is very high. I think, therefore, that there is no good reason for this divergence over fundamental questions concerning what Frege was about" (xi). Although I agree that Frege's standard of lucidity is very high, I do not think that such lucidity will, in general, preclude misunderstanding. For I believe the divergence that Dummett laments is attributable both to the influence Frege's writings have had on the history of analytic philosophy and to the subtlety of Frege's ideas.

[2]This is Paul Benacerraf's strategy in "Frege: The Last Logicist" (LL), where he investigates Frege's motivation. Benacerraf's conclusion is that Frege's project is primarily a mathematical project. Many of the arguments below are responses to Benacerraf's arguments. I discuss Benacerraf's paper more explicitly in "The Philosopher behind the Last Logicist."

project. For, were these his only motivations, Frege ought to have been content with available work by other mathematicians.

Frege's initial statement of his motivation appears to stem from a criticism of mathematical practice. He claims that mathematics, as it is practiced, is insufficiently rigorous. His overall project is an attempt to change this. As I will try to show, Frege was quite serious about this claim. He did believe that mathematical practice was too sloppy and that mathematicians ought to adopt new standards of rigor. Indeed, his *Begriffsschrift* introduces new standards of rigor which Frege believed mathematicians ought to adopt. However, it is important to remember, in this context, that his critique of mathematical practice is introduced as part of an argument that is intended to motivate the opening question of *Foundations*. And, with this in mind, it is revealing to examine this argument in detail, for it is not as obvious or natural as it looks. It is not clear that his criticism of standards of rigor in mathematics (or any other criticism of mathematical practice) really motivates the opening question of *Foundations*.

Frege introduces this question by noting the surprising lack of agreement among mathematicians in their characterizations of the number one. He says,

> Yet is it not a scandal that our science should be so unclear about the first and foremost among its objects, and one which is apparently so simple? Small hope, then, that we shall be able to say what number is. If a concept fundamental to a mighty science gives rise to difficulties, then it is surely an imperative task to investigate it more closely until those difficulties are overcome. (FA ii)

But it is not clear what the difficulties to which Frege alludes are. Presumably, if this criticism is to have impact, these difficulties must affect actual mathematical research. Thus, it should not be sufficient to show that different mathematicians have said different (and incompatible) things about what numbers are. The different claims about the nature of the numbers must result in differences concerning which proofs are accepted as legitimate or which arithmetical claims are accepted as truths. But Frege does not argue that there are such differences. He never brings up disagreements over proofs of theorems which could be traced to differences over the nature of the concept of number or the number one.

Of course, Frege's work is not typically read as an attempt to eliminate existing conflicts among mathematicians. A more likely interpretation of his mathematical worries is that they concern the possibility that arithmetic, and hence all mathematics based on it, might be inconsistent.[3] Two pieces of textual evidence seem to provide support for this view of the nature of Frege's worries. One of these is the following passage:

> Yet it must still be borne in mind that the rigour of the proof remains an illusion, even though no link be missing in the chain of our deductions, so long as the definitions are justified only as an afterthought, by our failing to come across any contradiction. By these methods we shall, at bottom, never have achieved more than an empirical certainty, and we must really face the possibility that we may still in the end encounter a contradiction which brings the whole edifice down in ruins. For this reason I have felt bound to go back rather further into the general logical foundations of our science than perhaps most mathematicians will consider necessary. (FA ix)

This remark might be taken as an argument that, in absence of a definition of the number one, there is no justification for thinking that arithmetic is consistent.[4] If Frege can argue convincingly for this, of course, then he seems to have provided his project with a compelling motivation.[5] However, it is by no means clear that Frege is suggesting in this passage that only rigorous definitions of the numbers will provide us with justification for thinking that arithmetic is consistent. In order to make a case for such a reading, it will be necessary to be clear about what "these methods" and "the whole edifice" are.

First, this passage introduces a worry about the consistency of arithmetic only if the methods in question are used almost universally in our justifications of arithmetical propositions.[6] Now if, when Frege says

[3]For instance, Benacerraf, who thinks Frege should be taken at his word when he says that his worries are mathematical, also claims (LL 23) that Frege was worried about the inconsistency of arithmetic.

[4]As Benacerraf does. See LL 23.

[5]This is not entirely uncontroversial. Philip Kitcher seems to believe that worries about consistency are not, on their own, sufficient to motivate mathematical research. In particular in chapter 10 of *The Nature of Mathematical Knowledge*, Kitcher argues that, at least in the history of analysis, worries about consistency were not sufficient to motivate mathematical research. However, I do not think Kitcher's argument warrants the conclusion that, if a mathematician decides there are worries about consistency in some field and proceeds to give proofs intended to show that these worries are unfounded, the project is unmathematical or not mathematically motivated. I think the most that can be said is that in such a case the mathematician's worries may lie outside the mainstream of mathematical research.

[6]Although Frege distinguishes symbols from what they express (see, e.g., FA 13) and from their meaning (significance, reference) and/or sense (*Bedeutung* and/or *Sinn*, see,

"by these methods," he is talking about the practice of not explicitly defining the fundamental notions of arithmetic, he certainly could be taken as voicing such a worry. But this is not what Frege is talking about. Frege begins the paragraph that ends with the passage quoted above by saying that he is departing from mathematical practice. He then characterizes this practice as follows:

> If a definition shows itself tractable when used in proofs, if no contradic-
> tions are anywhere encountered, and if connections are revealed between
> matters apparently remote from one another, thus leading to an advance
> in order and regularity, it is usual to regard the definition as sufficiently
> established. (FA ix)

This is the practice that Frege is criticizing throughout the paragraph in question. He thinks that a definition should only be introduced if it can be proved not to lead to inconsistencies. But Frege's criticism is not that mathematicians' definitions of the numbers are not adequately justified, but rather that they have no definitions at all.

Second, it is not clear that the edifice Frege means is arithmetic. Presumably, Frege is simply talking about any mathematical structure that is built up from inadequately justified definitions. Arithmetic is not built up from inadequately justified definitions, since it is not built up from definitions at all. Thus, Frege does not seem to be arguing in this passage that, in absence of a definition (or, at least, a uniform mathematical characterization) of the numbers, there is a real worry that arithmetic is inconsistent. There is other evidence that Frege did not

---

e.g., FA 106–113), he seems not to distinguish between sentences of arithmetic and what is expressed by these sentences. In his chapter 1 discussions of truths and proofs of arithmetic, Frege seems to use *Satz*, which Austin translates as 'proposition', and *Formel*, which Austin translates as 'formula', almost interchangeably. Frege speaks of proving numerical formulae as well as proving propositions. For instance, Frege says that we have to distinguish numerical formulae (*Zahlformeln*) such as $2 + 3 = 5$ from general laws (FA 5), and he talks about the proof of the proposition (*des Satzes*) $5 + 2 = 7$ (FA 10). (In the passages cited, the expressions '$2 + 3 = 5$' and '$5 + 2 = 7$' appear without quotation marks.) This blurring of the notions of sentence and what a sentence expresses is also in evidence in the final sections of *Foundations*.

The arguments from the chapter 1 sections with which I am concerned here could be rewritten in such a way as to eliminate use/mention confusions. For instance, one might rewrite Frege's heading "Are numerical formulae provable?" (*Sind die Zahlformeln beweisbar?*) as "Do numerical formulae express provable truths?" Frege's talk of proofs of numerical formulae could be replaced by talk of proofs of propositions that can be expressed by numerical formulae. However, in some cases, the result would be quite awkward and the resulting characterizations of Frege's arguments would not fit with the quotations from the text. I have chosen, therefore, to conform to Frege's expressions whenever the result will not cause serious confusion.

really have this worry. For, in the actual working out of his project, Frege never really seems seriously to consider the possibility that arithmetic might be inconsistent. When Frege introduces his mathematical motivations explicitly and in detail (in sections 1 and 2 of *Foundations*) he does not say that arithmetic might be inconsistent. And later in his career, after he had realized that his logical definitions had failed and after he had despaired of giving any logical definitions of the numbers, he did not consider the possibility that arithmetic might be inconsistent; rather, he inferred that there must be another way of finding the definitions of the numbers (see, in particular, the post-1924 writings in PW).

The second piece of textual evidence which seems to support the claim that Frege was worried about inconsistency comes from the explicit mathematical motivation Frege gives in sections 1 and 2 of *Foundations*. In section 1 he appeals to a tradition, in the history of mathematics, of demanding more and more rigorous proofs. He says, "The concepts of function, of continuity, of limit, and of infinity have been shown to stand in need of sharper definition" (FA 1). In section 2 he goes on to say, "Proceeding along these lines, we are bound eventually to come to the concept of Number and to the simplest propositions holding of positive whole numbers, which form the foundation of the whole of arithmetic" (FA 2).[7] On the surface, it may not seem that this has anything to do with a worry about consistency. After all, the mathematical tradition to which Frege appeals may simply be a tradition of defining all mathematical concepts. However, even if there were such a tradition in mathematics, Frege does not seem to be appealing to it. There is considerable evidence that Frege thinks it is appropriate to give reasons for wanting a definition of a term. First, his suggestion in the introduction to *Foundations* is that we should want a definition of the concept of number because this concept gives rise to difficulties (FA ii). Second, his claim, in section 1, that the concepts of function and so on have been *shown* to stand in need of sharper definition indicates that some justification may be required for asking for a definition of a mathematical concept. Furthermore, Frege never seems to advocate definition for its own sake, and he never suggests that all concepts, or even all mathematical concepts, should be defined. In fact, he suggests at one point (FA 5) that the concept of number might be indefinable. Indeed,

---

[7]The Austin translation contains the following footnote: "[*Anzahl*, i.e., cardinal number, cp. §4 n. I have always used "Number" to translate this and "number" for the more usual and general *Zahl*. Throughout most of the present work the distinction is not important, and Frege uses the two words almost indifferently.]"

throughout the first chapter of *Foundations*, it seems clear that Frege is undertaking the burden of showing that numbers stand in need of definition.

It is not obvious that Frege's demand for a definition of the number one and concept number should be viewed as part of the mathematical tradition. In order to see how Frege might have viewed it in this way, it will be necessary to examine how the demands for definitions of the concepts he mentions arose in mathematics. Frege's claim is that the demands for sharper definitions of the concepts of function, continuity, limit, and infinity arose from the desire for more rigorous proofs on the part of mathematicians. He suggests that, if proofs are to be as rigorous as possible, it is necessary to give precise definitions of the numbers as well. However, while Frege is right in saying that the mathematicians who wanted these definitions were after more rigor in mathematics, these were not demands for rigor for its own sake. In general, the demand for a definition of some concept arose when it was discovered that the concept, as it was commonly understood, supported results that were inconsistent with each other or prohibited solutions to important problems. This is the sense in which, if Frege's explicit mathematical motivation for defining the numbers is to be taken seriously, it must arise from worries about the consistency of arithmetic. To see this, it may help to consider a somewhat oversimplified account of some of the problems that arose with the mathematical concept of limit before it was given a precise definition.[8]

---

[8]The following account is a simplified version of that given by Morris Kline in *Mathematical Thought from Ancient to Modern Times*. My aim is not to give a satisfying account of the history of the notion of limit here, but rather to give the reader a sense of the difference between Frege's demand for a definition of the number one and earlier demands for a definition of 'limit'. The point is that the need felt by mathematicians for a definition of 'limit' was a response to an actual mathematical problem. Following Kline, I have suggested that certain inconsistencies forced mathematicians to worry about the notion of limit.

Kitcher has argued in *The Nature of Mathematical Knowledge*, chapter 10, that even the apparent inconsistencies in analysis were insufficient for motivating most mathematicians to worry about the notion of limit. The real push for a definition of 'limit' was motivated, according to Kitcher, by the realization that the absence of such a definition presented obstacles for solving other mathematical problems. On this story, the difference between Frege's project and the project of defining 'limit' is even more dramatic. Regardless of which account of the history of analysis one accepts, Frege's motivation for demanding that the number one be defined looks very different from the mathematical motivation for defining 'limit'.

Kitcher also notes the difference between the motivation for defining 'limit' and Frege's motivation for defining '1' in "Frege, Dedekind, and the Philosophy of Mathematics," 305–307, as does Gregory Currie in *Frege: An Introduction to His Philosophy*, 10–11.

The need for a definition of limit had its roots in the use of this concept for making sense of the notion of derivative. This notion was useful for solving a variety of problems. One way of thinking about the derivative is to consider the calculation of the velocity of an object at a particular time, assuming that we know its location at any point. Given an object that travels along a one-dimensional path over time, this problem can be represented by use of a graph of a function of one variable. Progress along the abscissa, or x-axis, indicates progress in time, and progress along the ordinate, or y-axis, movement in space. The velocity of the object, given such a representation, will be the rate of change of the function. If this representation is a straight line, the rate of change of the function will be constant: it will be the slope of the line. Thus, the velocity at any point can be calculated by considering the velocity over an interval containing the point. To calculate this on an interval, $[x, x_o]$, one divides the difference between the values of the function at x and x_o (i.e., distance traveled in the interval) by the difference between x and x_o (i.e., time elapsed):

$$\frac{f(x) - f(x_o)}{x - x_o}$$

The rate of change at a particular point will be the same. But it is more difficult to calculate the velocity if the rate of change varies. One strategy for making such a calculation is to consider an interval around the point which is so small that the rate of change is constant. Early definitions of the derivative, or rate of change, worked on the assumption that it was possible to calculate the rate of change of a function on an infinitely small interval around a point. This strategy involved use of the notion of infinitesimals, numbers that were infinitely small but not equal to zero. In the 1820s, Cauchy formulated a definition of the derivative which was supposed to avoid problems involved with the use of infinitesimals by using the concept of limit. Cauchy's definition of limit was as follows: "When the successive values attributed to a variable approach indefinitely a fixed value so as to end by differing from it by as little as one wishes, this last is called the limit of all the others."[9] The derivative of a function, f, at x_o was then defined to be the limit, as h approaches 0, of the result of dividing the difference between the values of f at x_o and $x_o + h$ by h. (Or:

[9]This is the translation published in Kline, *Mathematical Thought*, 951.

$$\lim_{h \to 0} \frac{f(x_0 + h) - f(x_0)}{h} \quad .)$$

This definition was as useful as previous definitions but avoided the problems with infinitesimals.

But Cauchy's definition of limit was not precise enough. In particular, it is not at all clear what it is for a sequence to approach a fixed value or for things to differ as little as one wishes. And this vagueness was by no means innocent. Cauchy's notion of limit was used to define not only the notion of derivative but also the notion of continuity. A great many mathematicians of Cauchy's time, Cauchy included, assumed that it followed from these definitions that a continuous function must be differentiable.[10] In fact, there were numerous "proofs" of this "theorem." A serious problem arose when Weierstrass was able to find examples of continuous functions that were nowhere differentiable. It was clear from this that the notion of limit was not well understood.

Weierstrass solved this problem by defining limit in simple arithmetic terms, and his definition survives today. The limit as x approaches x_0 of $f(x) = L$, according to this definition, if and only if, for any ϵ we choose, no matter how small, there is a δ such that whenever x is no farther from x_0 than δ, $f(x)$ is no farther than ϵ from L. In other words, we can find an x for which $f(x)$ comes as close as we want to L by making sure that x is close enough to x_0. How close we want $f(x)$ to be to L will determine how close x must be to x_0. In this way Cauchy's vague ideas of a sequence's approaching a fixed value and differing from it as little as one wishes can be made precise. And this precise definition, of course, resolved the problem. The examples of continuous nondifferentiable functions held up (i.e., the functions could be shown to be continuous and non-differentiable), and the "proofs" of the "theorem" that every continuous function is differentiable did not hold up.

This sort of account is not peculiar to the history of the notion of limit. Similar stories can be told about the other mathematical concepts that Frege identifies as having been shown to stand in need of sharper definition. The problem with Cauchy's definition was not that it did not satisfy some abstract standard of rigor. The vagueness in his definition represented real confusion, and this confusion manifested itself, ultimately, in an apparent inconsistency. Weierstrass's precise definition of the concept of limit was needed in order to solve problems that arose

[10]But not Bolzano. See Kline 955.

in mathematical research. Frege's claim is that, proceeding along the lines that led mathematicians to demand definitions for the concepts of limit and related concepts we will come to see that there is a need for a definition of the number one. If this claim is to be taken seriously, the burden is on Frege to show that current understanding of the number one will present actual problems for mathematical research.

It is difficult to know how seriously this motivation should be taken. Frege never undertakes the burden of showing that, eventually, problems in actual mathematical research will be traceable to confusions concerning the number one. Furthermore, as I indicated earlier, throughout his career Frege seems to rely on the assumption that arithmetic is consistent. If the mathematical motivation Frege provides in sections 1 and 2 is to be taken seriously, there seems, at the very least, to be a tension between these sections and the rest of his work. It is also worth noting that this motivation is not likely to appeal to mathematicians. Arithmetic has a much longer history than analysis and has never exhibited the sort of difficulties that plagued analysis. It is not very plausible to suppose that there are problems with arithmetic. Furthermore, the history of mathematics seems to indicate that, should arithmetic be defective, the appropriate time to deal with its defects will be after these defects have created problems for mathematical research.

Frege's point might be read with a somewhat different emphasis. As I indicated earlier, Frege is critical of actual mathematical practice. And the history of the definition of the concept of limit may well exemplify the practices of which Frege is critical. After all, the problems that arose did result from a lack of rigor in mathematics. Frege may be suggesting that a serious increase in rigor now will ensure that such situations do not arise in the future.[11] In order to ensure this, it will be necessary to prove everything provable (including obvious mathematical truths) and define everything definable (including what may seem to be primitive mathematical notions). It does not seem unreasonable to suppose that, were all these questions answered, were such a project carried out and

---

[11]Dummett has suggested in *Frege: Philosophy of Language*, xvii, that the purpose of Frege's attempt to construct a formal system in which mathematical proofs could be formulated was to prevent errors in proofs from arising. In this sense, some of Frege's work could be viewed as proceeding along the lines that led mathematicians to demand a definition of the concept of limit. However, it is more difficult to make this argument for Frege's demand for definitions of the numbers. Dummett does not say that this demand was in accord with the mathematical tradition.

the results accepted by mathematicians, no more problems of incompatible theorems would arise.[12]

How would such a project be carried out? If Frege is not suggesting that the proving and defining go on forever, there must be some means of determining what is definable or provable and there must be some primitive notions that are usable in definitions, some primitive truths that can serve as the basis for proofs. From a mathematical point of view, arithmetic would seem to be the right place to start. The development of analysis is characterized by its arithmetization—by the replacement of such mysterious terms as 'limit' with precise definitions in arithmetical terms. But, from a mathematical point of view, arithmetic may also be the right place to stop. In an important sense, the numbers and laws of arithmetic seem to be primitive. There are no other, more obviously primitive, mathematical terms that could be used to define the numbers. Why should definitions of the numbers from new, less familiar, and less obviously primitive terms provide further justification for the consistency of arithmetic?

Of course, even if most of Frege's explicit mathematical motivations are not natural at all, it does not follow that Frege's project had no mathematical motivation. In the last paragraph of section 2 Frege gives what looks to be a perfectly acceptable and, indeed, obvious mathematical motivation. He says, "The aim of proof is, in fact, not merely to place the truth of a proposition beyond all doubt, but also to afford us insight into the dependence of truths upon one another." It is important to remember, after all, that new proofs (even new proofs that do not solve any antecedently worrisome mathematical questions) are of mathematical interest. A series of proofs may well be evaluated not on grounds of what problems they solve but simply by how interesting they are or by

[12]Frege describes his project in this way on a number of occasions. He says, in "On Sense and Meaning," "A logically perfect language [Begriffsschrift] should satisfy the conditions, that every expression grammatically well constructed as a proper name out of signs already introduced shall in fact designate an object, and that no new sign shall be introduced as a proper name without being secured a meaning. . . . The history of mathematics supplies errors which have arisen in this way. . . . It is therefore by no means unimportant to eliminate the source of these mistakes, at least in science, once and for all" (CP 169). He also describes his project in this way in PW 207–241 and CP 300–304. Of course it may very well be unreasonable to suppose that such a project could be carried out. But it should be clear that, at least for mathematics, Frege is committed to carrying out such a project. And, although I have suggested that this might be taken as a mathematical motivation of Frege's project, it is worth noting that this motivation has a strong philosophical flavor.

how much insight they seem to provide. In this sense, Frege's project did have a mathematical motivation. But this is really not to say anything more about Frege's motivation than that it resulted in his proving theorems. We already know that Frege proved theorems. But there is reason for thinking this explanation does not exhaust Frege's motivation.

Had Frege's motivational remarks appeared, on their own, in the introduction to *Basic Laws*, and had his *Foundations* never appeared at all, it would be easier to understand his project and motivation as mathematical and to evaluate the interest of his work by considering the mathematical interest of his proofs. But the very existence of *Foundations* suggests that Frege's project is more ambitious than this. Were the interest of Frege's project to lie solely in the mathematical interest of his proofs, the existence of a lengthy discursive discussion of his project would seem to require some explanation. Why should he not have been content with a few prefatory remarks in *Basic Laws*? One possible explanation lies in the novelty of Frege's question. There were no ready-made tools available to Frege from which proofs of the basic truths of arithmetic could be derived. For, since Frege wanted to prove all the unproved truths of arithmetic, no tools of arithmetic or truths of arithmetic could provide the basis for Frege's proofs. And there was no set of truths which was generally regarded by mathematicians as underlying the truths of arithmetic—arithmetic was thought to be the most basic part of mathematics. Frege is generally regarded as having invented and used tools provided by his formal logic and set theory. The obvious strategy, then, is to try to read *Foundations* as an attempt to make the nature of these tools clear.

But this sort of reading will not work. For none of these tools is first explicated in *Foundations*. Frege's formal logic had been developed in his *Begriffsschrift* and there is no attempt in *Foundations* to make it any clearer. Furthermore, the tools of set theory were available in Cantor's work (to which Frege himself alludes), and Frege does not even discuss set theory in *Foundations*. Thus, had Frege's project been to define the number one and concept of number in terms of logic and set theory and to use these definitions to prove all truths of arithmetic, it would have sufficed to add a brief discussion of Cantor's set theory to his *Basic Laws*. There would surely be no important role for *Foundations* to play. The very fact that *Foundations* was written constitutes some evidence that Frege did not see himself as engaged in the project just described.

And there is even better evidence than this. In *Foundations*, Frege explicitly denies that this is what he is doing. He claims (FA 38) not to understand the notion of set and says that he would not recognize a definition of number in terms of set as an answer to his questions. Indeed, Frege explicitly discusses (BLA 4) Richard Dedekind's essay *Was sind and was sollen die Zahlen?* in which the numbers are defined from the terms 'system'—which is used, more or less, for the mathematical notion of set—and 'transformation'. Frege objects that Dedekind's notion of system is not a logical notion and is not reduced to acknowledged logical notions.[13] Frege sees set theory as being in as much need of foundation as number theory. It is significant that, just as there is no mathematical reason for not taking the concept of number as primitive, there is no mathematical reason for not taking the concept of set as primitive. The criteria on which set theory and arithmetic require foundations do not seem to be mathematical criteria.

If the aims and pretensions of Frege's project are viewed as solely mathematical, it is not obvious that his *Foundations* has a role to play. Also, if Frege's project is to be regarded as having a compelling motivation, certain questions concerning the nature of his understanding of primitiveness, definition, and proof must be answered. And it is clear that there is something very unmathematical about his understanding of these notions. Thus, it seems that there will be significant gaps in any purely mathematical account of Frege's project. Since Frege claims to have a philosophical as well as a mathematical motivation, it seems only reasonable, at this point, to turn to his discussion of this philosophical motivation for answers. It is important to note, however, that the mere fact that he put forth a philosophical motivation for his project does not set Frege's work apart from that of most mathematicians who are engaged in purely mathematical projects. It is not unheard of for a mathematical work to be prefaced by an apparently philosophical discussion. Dedekind's work is a good example.

In the preface to *Was sind und was sollen die Zahlen?* Dedekind appears, as Frege does, to give a philosophical motivation for his project of defining the numbers, which is accomplished, as Frege's project is, in a book of proofs. Dedekind claims, in his preface, that he means to show

[13]Of course, given the novelty of Frege's logic, it may not be clear that his logical notions should count as "acknowledged logical notions." However, Frege did believe that his notion of the extension of a concept would, upon its introduction, be immediately recognized as a logical notion. The special logical nature of Frege's notion of the extension of a concept will become clearer in chapter 2 herein, when Frege's understanding of logic is examined in detail.

that numbers are "free creations of the human mind."[14] But once one is immersed in Dedekind's text this remark about numbers and the human mind is easily forgotten. It is cashed out only in an informal explanation of the use of his term 'system' (set)—an explanation that, ultimately, plays no important role in determining the mathematical character of systems or in a proof of the existence of infinite systems.[15] No use of this remark includes any substantive philosophical discussion or is clearly based on a substantive philosophical view. Although Dedekind uses the fact that different objects can be associated in the mind to explain his notion of objects being elements of a system, this does not distinguish his notion of system from any other notion of set.

His proof of Theorem 66 ("There exist infinite systems," EN 64) may look more philosophical. For it depends on the assumption that, for every possible object of thought, s, there is the distinct possible object of thought that s can be an object of thought. Dedekind uses this assumption to show how to give a mathematical definition of a provably infinite set of possible objects of thought. The connection between thought and mathematics seems to be exploited here as the briefest possible defense of Dedekind's use of the notion of infinity. But Dedekind does not discuss the many obvious difficulties inherent in the notion of a realm of possible objects of thought. The lack of such a discussion makes it seem unlikely that there is much substance to Dedekind's claim that numbers are free creations of the human mind or that this claim provides a serious motivation for his work. While it is easy to evaluate Dedekind's work as a work of mathematics, it is very difficult to say in what sense his proofs provide an answer to the philosophical questions with which they are introduced.[16]

---

[14]In *Essays on Number* (EN), 31. All further references to this work in this chapter will be made parenthetically in the text.

[15]I am indebted to Tom Ricketts for pointing out that Dedekind does actually talk about the realm of thoughts in his argument for the existence of infinite systems (sets).

[16]Kitcher has argued, in "Frege, Dedekind, and the Philosophy of Mathematics," that Dedekind's claim that numbers are creations of the human mind is serious because it is cashed out, mathematically, in his "creation" of the real numbers via partitions of the rationals. But, while this notion of creation may have functioned as a useful heuristic for Dedekind cuts, it is not clear that the end result is significantly different from any other mathematical definition. Dedekind says: "Whenever, then, we have to do with a cut (A_1, A_2) produced by no rational number, we create a new, an *irrational* number α, which we regard as completely defined by this cut (A_1, A_2)" (EN 15). But nothing more is said about the mind here. In fact, Dedekind then goes on to show that his characterization of the irrationals provides identity criteria. Thus, it is not clear that there is any real substance to Dedekind's views about the mind.

In Chapter 8 of *Frege, Dedekind, and Peano on the Foundations of Arithmetic*, Donald Gillies also characterizes Dedekind as having primarily mathematical rather than philosophical

I will try to show that Frege's project is philosophical in a much deeper sense. It is not simply that more substantive philosophical views can be extracted from Frege's *Foundations* than from the preface to Dedekind's essay. Philosophical considerations rather than mathematical considerations lead Frege to reject many apparently similar mathematical works (including Dedekind's) as providing answers to his questions. Frege's own proofs look very different from the proofs in those works. A serious account of the philosophical motivations of Frege's project will show how philosophical considerations dictate many of the mathematical details of Frege's work.

In sections 3 and 4 of *Foundations*, Frege introduces the philosophical questions his project is designed to answer. The remainder of chapter 1 of *Foundations* is devoted to arguments against the views of other philosophers, particularly Kant and Mill. If the arguments in these sections are understood as attempts to refute the views of the philosophers discussed, they can only be regarded as failures. For Frege seems to make very little effort to understand the views he discusses, and many of his arguments against them look naive and shallow. However, there is a more interesting way to read these sections. Frege can be viewed not as trying to refute all other philosophical views concerning the nature of number, but rather as trying to argue that none of these views will provide answers to his philosophical questions. In order to make the nature of these philosophical questions clear, it will be useful to examine much of chapter 1 of *Foundations* in detail.

THE PROVABILITY OF NUMERICAL FORMULAE

Sections 5 and 6

Frege begins his discussion of the provability of the numerical formulae with an objection to Kant. It is probably worthwhile to point out that

---

interests. Gillies notes that, although Russell ridiculed Dedekind's "creation" of the real numbers via partitions of the rationals as having "the advantages of theft over honest toil" (in *Introduction to Mathematical Philosophy*, 71), all Dedekind need have done to satisfy Russell would have been to define each real as a set of rationals. Russell was not objecting to Dedekind's use of notions having to do with the mind but only to the lack of the existence proof that explicit definitions would allow.

Kitcher also wants to attribute a serious and defensible philosophical view to Dedekind. But it is interesting to note that the elaboration and defense of this view which Kitcher discusses are Kitcher's, not Dedekind's. Kitcher admits that Dedekind did not address the obvious difficulties inherent in the view. The upshot of Kitcher's exposition of Dedekind seems to be that Dedekind made a number of suggestive remarks in his essays, some of which may be of interest to contemporary philosophers who worry about mathematics.

Frege does not begin his criticism of Kant's view with a lengthy discussion of what exactly Kant's view was. In fact, Frege makes only two identifiable claims about Kant in these sections. First, Frege claims that on Kant's view arithmetical truths are fundamental, unprovable, and known through intuition. Second, Frege mentions that Kant appeals to our intuition of fingers or points in order to explain the justification of truths of arithmetic. Neither of these claims is fleshed out. It might seem entirely reasonable, at this point, to investigate Kant's actual views and discuss the legitimacy of Frege's criticism of these views. But to do so would, in the end, be counterproductive. Frege does not raise objections to a well-worked-out view. His arguments are meant simply to convince his readers that the claim that intuition directly justifies all numerical formulae is implausible.

Frege has two sorts of arguments. In his more straightforward arguments, he seems simply to be saying why he does not find Kant's explanation of our knowledge of the numerical formulae satisfying. Kant, he says, appeals to our intuition of fingers or points in order to explain the justification of numerical formulae. But, Frege says (FA 6), "even 10 fingers can, in different arrangements, give rise to very different intuitions." Frege's other sort of argument has to do with problems resulting from Kant's taking certain numerical formulae not to be ultimately justified by proof. Frege asks (FA 6) if it is really self-evident that $135664 + 37863 = 173527$. Even Kant, he says, thinks that it is not, and Kant takes this non-self-evidence as an indication that such propositions are synthetic. Frege takes this to illustrate the unsatisfactory nature of Kant's view, for since such formulae cannot be proved on this view, it is not clear how they are to be justified. Frege sees the non-self-evidence of such formulae as an indication that they are provable.

Frege goes on to consider an alternate position for Kant. He supposes that Kant might have acknowledged that some, but not all, numerical formulae require proof. That is, perhaps propositions holding of large numbers are provable, but propositions holding of small numbers are immediately self-evident through intuition.[17] But Frege argues that this

---

[17]This talk of a proposition's "holding of" something is intended to fit into Frege's locutions. For instance, "the simplest propositions holding of positive whole numbers" (*die von positiven ganzen Zahlen geltenden einfachsten Sätze*, FA 2) and "Empirical propositions hold good of what is physically or psychologically actual" (*Die Erfahrungssätze gelten für die physische oder psychologische Wirklichkeit*, FA 20). My aim, in using this somewhat awkward locution, is to avoid talking of a proposition's being *about* something—since Frege later seems to believe that "aboutness" is a feature of a sentence, not of the thought expressed (see, e.g., CP 188).

is not convincing. For, if intuition provides us with immediate access to small, but not large, numbers, this constitutes a fundamental difference between large and small numbers. And if such a fundamental difference exists, the distinction between large and small numbers should be obvious. But it is not. Not only is there no clear-cut difference between large and small numbers, but it is not even clear where the line between large and small should be drawn. This, of course, does not constitute a devastating argument against Kant's views, nor is it meant to do so. Rather, Frege takes all this as evidence for his claim that we just do not seem to have the intuitions of numbers on which, according to Kant, our knowledge of arithmetic is based.

Since Frege has claimed, in response to Kant, that numerical formulae are provable and require proof, he must show that there are proofs of at least some numerical formulae. Proving some numerical formulae is, in any case, a central task for Frege. In his description of how to determine whether the truths of arithmetic are analytic or synthetic, a priori or a posteriori, Frege says that we must prove these truths from primitive truths. Thus, Frege must isolate those truths of arithmetic which are not provable in arithmetic and either conclude that they are primitive or prove them from truths that are primitive.

In section 6 Frege begins to discuss proofs of numerical formulae. In this section Frege exhibits a method, which he attributes to Leibniz, for proving all numerical formulae that contain numbers greater than one. First, Frege claims, all numbers can be defined from their predecessors using the notion of increase-by-one. That is, Frege envisions the definitions of the numbers as follows: 2 is defined as 1 and 1—or 1 increased by 1, or 1 + 1. Now that 2 has been defined, 3 can be defined as 2 and 1—or 2 increased by 1. Of course to say this is simply to abbreviate the definition of 3 as 1 increased by 1 and then increased by 1 again, or (1 + 1) + 1. In this way, all the numbers can be defined from one and increase-by-one. Frege says, "Through such definitions, we reduce the whole infinite set of numbers to the number one and increase-by-one, and every one of the infinitely many numerical formulae can be proved from a few general propositions" (FA 8). Frege also gives an example of how this method works. His example is a modified version of a proof, which he attributes to Leibniz, of the claim that 2 + 2 = 4. First, 2, 3, and 4 are defined from one and increase-by-one, as described above, and then these definitions are used, along with the associative law for addition (whose omission, Frege says, was an error in Leibniz's proof), to prove that 2 + 2 = 4. The proof Frege attributes to Leibniz is as follows:

Definitions. (1) 2 is 1 and 1
(2) 3 is 2 and 1
(3) 4 is 3 and 1

Axiom. If equals be substituted for equals, the equality remains.

Proof. $2 + 2 = 2 + 1 + 1$ (by def. 1) $= 3 + 1$ (by def. 2) $= 4$ (by def. 3). $\therefore 2 + 2 = 4$ (by the axiom). (FA 7)

This proof, Frege notes, does not quite work. The definitions do not allow us to calculate $2 + 1 + 1$. We can calculate only $(2 + 1) + 1$ and $2 + (1 + 1)$. In fact, $2 + 2 = 2 + (1 + 1)$ and $3 + 1 = (2 + 1) + 1$. In order to conclude that $2 + 2 = 3 + 1$, then, it is necessary to show that $2 + (1 + 1) = (2 + 1) + 1$. This is an instance of the associative law of addition. The proof that $2 + 2 = 4$ should actually read:

$2 + 2 = 2 + (1 + 1)$ (by def. 1) $= (2 + 1) + 1$ (by associative law) $= 3 + 1$ (by def. 2) $= 4$ (by def. 3).

In this way, Frege claims, every formula of addition can be proved and the infinite set of numbers can be reduced to one and increase-by-one. Furthermore, by use of the definitions, any numerical formula in which numbers greater than one occur can be proved from a numerical formula in which only the number one occurs.

These proofs of the numerical formulae from formulae in which only the number one occurs simplify Frege's task of finding the primitive truths on which arithmetic depends. Indeed, it may seem, at this point, that Frege need only account for the justification of the definitions and of numerical formulae in which only the numeral 1 appears. However, the omission from Leibniz's proof illustrates the need to account for another sort of truth of arithmetic, the general laws. Frege argues that attempts to obtain the general laws of arithmetic by means of definitions have failed. Thus, Frege must account for the justification of three sorts of truths of arithmetic: those expressed by numerical formulae that contain no names for numbers greater than one, definitions of larger numbers that make use only of one and increase-by-one, and general laws of arithmetic. Frege believes that all truths of arithmetic can be given proofs in which only truths from these three groups are needed. In order to answer Frege's question about whether the truths of arithmetic are analytic or synthetic, a priori or a posteriori, it is necessary and sufficient to determine the status of propositions from these three groups.

It should be noted that this also suggests a response to one of Frege's criticisms of Kant (although Frege does not consider this answer). Frege has argued that, given Kant's views, there ought to be a principled way of distinguishing between large and small numbers and that there is none. But in section 6 of *Foundations*, Frege himself shows that there is a principled way of distinguishing between large and small. If all numbers can be defined from one and increase-by-one, then surely the line between large and small should be drawn between the numbers two and one. That is, all numbers larger than one should be viewed as large numbers. It is consistent with Frege's view, then, that we could have immediate access through intuition to the facts about the numbers zero and one and that all other arithmetic knowledge could be derived via proofs from these and the general laws. It is by no means clear, of course, that Kant would want to follow this strategy (or even, for that matter, that the views described by Frege as Kant's views actually were Kant's views). However, it should be clear that the position Frege takes to be the Kantian position has not been conclusively refuted by Frege's arguments. Frege's only real objection to this position is that he finds it unsatisfying. The purpose of his argument is to persuade his readers that it is unsatisfying.

In the last few sections Frege has argued that the justifications of all truths of arithmetic depend on propositions from the following categories: definitions of larger numbers using only one and increase-by-one; propositions expressed by numerical formulae in which no names for numbers greater than one appear; and general laws of arithmetic. And Frege's goal is to convince his readers that no propositions of arithmetic are primitive—that is, unprovable *and* immediately identifiable as either logical truths or truths of some special science. In order to accomplish this it suffices to argue that no propositions in the three categories mentioned are primitive. There is no problem with the definitions of larger numbers in terms of one and increase-by-one. For these are unprovable but also are neither truths of logic nor truths of any special science—they can be regarded as abbreviations. The use of these definitions in a proof of some proposition will not in any way determine the status of the proposition. The second category is that of propositions that are expressed by numerical formulae in which no names for numbers greater than one appear. Frege's discussion of Kant's view is designed to convince us that numerical formulae are not obviously synthetic a priori. But the status of a primitive truth must be obvious. A primitive synthetic a priori truth must be obviously synthetic a priori. Thus, if any numerical formulae are primitive, they will have to be primitive synthetic a posteriori truths. (Needless to say, no numerical formulae are primitive logical truths.[18]) Frege's next step, then, will be to argue that no numerical formulae express primitive synthetic a posteriori truths. Once he completes this argument, he will return to the third, and final, category of basic arithmetic truths—the general laws. The final part of his argument that no truths of arithmetic are primitive will consist of an argument that the general laws of arithmetic are neither clearly synthetic a priori nor clearly synthetic a posteriori.

Sections 7 and 8

In sections 7 and 8 of *Foundations*, Frege discusses Mill's views. The tenor of this discussion is very different from that of his discussion of Kant's views. Although Frege invariably treats Kant with respect, the tone of his discussions of Mill and Mill's arguments borders on contempt. Are Mill's arguments as silly as Frege indicates? Some of the arguments Frege attributes to Mill do seem silly; thus, it is important to remember that Frege is not the most generous reader of other philosophers' views. Where Mill's arguments are concerned, however, the problem is less Frege's lack of charity than the total lack of common ground between Frege's views and Mill's. For although Frege and Mill use many of the same basic expository terms, it is clear that they understand these terms differently. Frege indicates that his aim is to determine whether the truths of arithmetic are analytic or synthetic, a priori, or a posteriori. He provides us with new definitions of these terms and an account of what would establish the truths of arithmetic as analytic, synthetic, a priori or a posteriori. Mill does not use these terms in Frege's sense. More significant, the assumptions that run through Frege's account already amount to a denial of Mill's view—once they are accepted, nothing Mill says can count as an answer to any of Frege's questions. It almost does not matter that Frege takes Mill's views out of context, since the views that constitute their context have already been dismissed as absurd. Thus, Frege's arguments against Mill, in chapter 1

---

[18]This is not to say that no numerical formula can be a logical truth. For example, it is a logical truth that $1 = 1$. On the other hand, this formula should not be primitive because, as Frege shows in *Basic Laws*, it is provable from a logical law (Basic Law III).

of *Foundations*, are not especially interesting if they are read as a refutation of Mill's actual view. My aim, in discussing these arguments, will be neither to determine whether or not Mill's view provides an alternative to Frege's nor to determine whether or not Frege's arguments work against Mill's real views. I will be interested primarily in determining what these arguments can tell us about Frege's picture.

Mill's view, as Frege represents it, is that arithmetical truths are empirical (synthetic a posteriori) and numerical formulae assert observed matters of fact. Frege presents a barrage of arguments against the claim that numerical formulae assert observed matters of fact, the most central of which seem to fall into two categories. The first category consists of Frege's indicating that he simply does not see how empirical facts could be asserted by numerical formulae or how our observations justify numerical formulae. This sort of objection is very like Frege's central (and, arguably, unconvincing) objection to Kant. However, this is more devastating to Mill's views than it was to Kant's. For, if numbers are simply taken to be primitive objects known through intuition (as, Frege admits, points are), there is always a response to Frege's objection. A Kantian could simply reply, "but I do intuit the numbers." But Mill can give no such reply. Mill cannot claim that numbers are physical objects that are seen (or otherwise sensed).

Of course, Mill's claim that numerical formulae are assertions about the empirical world does not entail the claim that numbers are empirical objects. There is another obvious strategy for explaining the sense in which numerical formulae might be assertions about the empirical world. The mathematical notion of limit was defined by a precise statement of what claims limit statements made about sequences of numbers. Mill could attempt to define the numbers by making precise, in non-numerical terms, the claims numerical formulae make about the world. Should Mill succeed, he might have a response to Frege. But Mill does not provide enough information for following this strategy—he does not really tell us what claims about the world are made by numerical formulae, and it is not obvious how these formulae might be viewed as claims about the world. Among the numerical formulae that, according to Mill, make claims about the world are such definitions as "$3 = 2 + 1$". But Frege argues that we cannot, in general, view such definitions as observed facts. For "what in the world can be the observed fact, or the physical fact . . . which is asserted in the definition of the number 777864?" (FA 9). This might look like an objection Frege raises to Kant's view. Frege suggested that, while it might seem plausible that intuition could justify arithmetical truths involving small numbers, it does not seem plausible that arithmetical truths involving large numbers could be justified in this way. As I noted earlier, this is an objection to which Kant could respond. For it is open to Kant to claim that intuition justifies truths about one and increase-by-one directly. Since all proofs of numerical formulae containing larger numbers will be dependent on facts about one and increase-by-one, these proofs must depend, ultimately, on truths known by intuition. Could Mill, similarly, argue that definitions of smaller numbers assert observed physical facts, while the definitions of larger numbers are to be regarded as generalizations?

Frege blocks this move by arguing that even definitions of small numbers cannot be regarded as assertions about the physical world. For he claims that even Mill's seemingly more plausible example of a physical fact asserted by a definition of the number three is unsatisfactory. Frege describes this example as follows:

> that collections of objects exist which, while they impress the senses thus, $^o o^o$, may be separated into two parts, thus, oo o. (FA 9)

This is unsatisfactory because, if this understanding of the sense in which the definition of three asserts physical facts is truly to explain what the number three is (or what the relationship between 1, 2, and 3 is), then there is no way of explaining how "3" can be used in descriptions of nonphysical phenomena.

> From this we can see that it is really incorrect to speak of three strokes when the clock strikes three, or to call sweet, sour, and bitter three sensations of taste. . . . For none of these is a parcel which ever impresses the senses thus, $^o o^o$. (FA 9–10)

Thus, Mill's explanation will not be adequate to explain the justification of all statements in which numbers are used. It should, incidentally, be noted for future reference that Frege has thereby set out a criterion of adequacy for definitions of the numbers. The definitions must be usable in explanations of justifications of all statements in which number words appear. An adequate definition of the number one, for instance, must be usable in accounts of the justification of such claims as that Frege has one horse as well as in the propositions of arithmetic (e.g., $0 \neq 1$).

Frege has a second sort of objection to Mill's proposed account. According to Frege, it follows by Leibniz's proof that numerical formulae are properly justified by proofs that rely only on definitions using the number one and increase-by-one along with general laws of number. Thus, Mill has the additional burden of showing that these justifications must rely on observed matters of fact. For, otherwise, the numerical formulae will not have been shown to be synthetic a posteriori. Frege says, about numerical calculations,

> Now according to MILL "the calculations do not follow from the definition itself but from the observed matter of fact." But at what point then, in the proof given above of the proposition 2 + 2 = 4, ought LEIBNIZ to have appealed to the fact in question? MILL omits to point out the gap in the proof, although he gives himself a precisely analogous proof of the proposition 5 + 2 = 7. (FA 10)

Frege is referring, in this passage, to the proof he discusses in section 6. He argues in section 6 that the gap in Leibniz's proof is a hidden application of the associative law. But, Frege notes here, Mill seems unaware of this gap. In particular, Mill does not exploit this gap when he argues that observed matters of fact play a role in justification of arithmetical truths. Thus, it seems that the only candidates for observed matters of fact which might play a role in the proof will be the definitions. If Frege's original objection to taking such definitions as '2 = 1 + 1' as physical facts holds up, then no observed matters of fact appear in the proof.

Mill could respond by claiming that the general law used is an empirical fact (although, as Frege notes, he does not seem to have recognized that this law is necessary for the proof). Were this true, however, the laws would have to be generalizations over numerical formulae containing numbers other than one. Thus, once again, as Frege says, "MILL's theory must necessarily lead to the demand that a fact should be observed specially for each number" (FA 10). That is, the definitions themselves would have to be observed matters of fact. And Frege has already argued that this is absurd.

ARE THE LAWS OF ARITHMETIC INDUCTIVE TRUTHS?

Sections 9–11

Frege has argued that the fundamental numerical formulae are neither clearly empirical nor clearly known immediately by intuition. This argument serves two purposes. First, it constitutes an argument for the claim that no numerical formulae are primitive truths. This is a first step toward arguing that no truths of arithmetic are primitive. Second, it lends plausibility to the claim that arithmetical truths might be analytic. In order to complete both of these lines of argument, Frege must also argue that empirical facts and intuitions do not obviously enter into the justification of the general laws of number. He addresses part of this in sections 9–11, where he argues that the laws of arithmetic are not inductively known. Although he gives a number of arguments, the central argument concerns our use of the general laws of number.

Frege's aim is to investigate the ultimate grounds of the justification of arithmetic. In order to accomplish this, it must be possible to show how the general laws of number can be justified from primitive truths. These laws cannot be justified inductively without using numerical formulae in the justification. Frege has already argued that the numerical formulae are not primitive truths. Thus, the justification of the general laws from primitive truths must also contain justifications of the numerical formulae used. That is, the justification of the general laws of number from primitive truths would begin with the justification of a series of numerical formulae from primitive truths followed by the inductive justification of the laws from the numerical formulae. But what would the justification of the numerical formulae be like? Frege has also already argued that the general laws of number are necessary for the justification of the numerical formulae. The problem with this is not simply that if the general laws of number must be inductively justified, their justification will be circular. Rather, the problem is that there is no way to get a straightforward justification of the general laws of number from the primitive truths that underlie them. Frege's assumption is that all scientific truths can be justified from primitive truths. It follows, on this assumption, that the ultimate justification of the general laws of number is not inductive.

ARE THE LAWS OF ARITHMETIC SYNTHETIC A PRIORI OR ANALYTIC?

Sections 12–14

Frege believes he has established, at this point in *Foundations*, that arithmetical truths cannot be shown to be synthetic a posteriori. He also has argued that the numerical formulae are not primitive synthetic a priori truths. Thus, arithmetical truths would fail, in an obvious way, to be analytic only if the general laws are synthetic a priori, that is (according to Frege), known through intuition. In sections 12–14, Frege argues that the laws of arithmetic are not synthetic a priori. Once more his strategy is to criticize Kant and, once more, his arguments are not conclusive. The general laws of arithmetic are simply truths that hold generally of the numbers. Thus, determining the ultimate justification of the general laws of arithmetic amounts to accounting for how we know things about the concept of number. Were these laws known by intuition, the knowledge of the concept of number must come from intuition. As before, Frege's argument against Kant's view seems to be that he just cannot find such knowledge in himself.

Frege also makes two specific criticisms of Kant's position in these sections. He accepts Kant's claim that geometry is synthetic a priori, but he argues that our knowledge of arithmetic differs from our knowledge of geometry in two ways. First,

> one geometrical point, considered by itself, cannot be distinguished in any way from any other; the same applies to lines and planes. . . . In geometry, therefore, it is quite intelligible that general propositions should be derived from intuition; the points or lines or planes which we intuite are not really particular at all, which is what enables them to stand as representatives of the whole of their kind. But with numbers it is different; each number has its own peculiarities. (FA 19–20)

That is, the strategy for arguing that the truths of geometry are synthetic a priori will not work for the general laws of arithmetic. Frege's second argument is that we can leave behind intuition—and geometry—in conceptual thought, but not arithmetic. He says that if we try to deny any fundamental proposition of arithmetic (FA 21), then "even to think at all seems no longer possible." This second argument is especially significant because it is more than an objection to Kant. This argument also provides direct grounds for thinking that arithmetic is analytic. Frege adds:

> The truths of arithmetic govern all that is numerable. This is the widest domain of all; for to it belongs not only the actual, not only the intuitable, but everything thinkable. Should not the laws of number, then, be connected very intimately with the laws of thought? (FA 21)

If the truths of arithmetic govern everything thinkable, it seems odd to take them as truths of any special science, for they must be applicable beyond that science. This is an independent reason for thinking that the truths of arithmetic are analytic.

It may be helpful, at this point, to review the general structure of the argument in the preceding pages. I have tried to show that there are two intertwined lines of argument in chapter 1 of *Foundations*. The explicit argument seems to be that arithmetic is not synthetic. Frege begins by providing criteria of adequacy for determining whether arithmetic is analytic or synthetic. He then tries to show that Kant's and Mill's attempts to show that arithmetic is synthetic do not meet these criteria. This task is completed in section 14. Of course, even if Frege is entirely successful, it does not follow that arithmetic is not synthetic. It only follows that Kant and Mill have not shown that arithmetic is synthetic. What would such a conclusion do for Frege? An obvious motive for providing convincing arguments against the more compelling theories on which arithmetical truth is synthetic is to convince his readers to consider seriously his attempt to show that arithmetic is analytic. But Frege's argument has more structure than this account of his motivation might indicate.

The real burden of Frege's argument is to show that the truths of arithmetic cannot be primitive. On Frege's understanding of primitiveness, primitive truths must be unprovable and the status of primitive truths (i.e., whether they are analytic or synthetic, a priori or a posteriori) must be obvious. But, except for instances of logical laws, which are provable, no truths of arithmetic are obviously logical (analytic) truths. Consequently, if some truths of arithmetic are primitive, they must be synthetic and it must be obvious that they are synthetic. His criticisms of Kant's and Mill's views are meant to show that the truths of arithmetic are neither obviously synthetic a priori nor obviously synthetic a posteriori. Thus, in criticizing the most plausible ways of taking arithmetic to be synthetic (Kant's and Mill's), Frege is also criticizing the claim that some truths of arithmetic are primitive. Furthermore, it would seem, then, that in order to answer Frege's question about the ultimate grounds of the justification of the truths of arithmetic, the concepts involved in arithmetical truths must be defined. In this way, not only has Frege motivated his search for an answer to the question about the status of the truths of arithmetic, he has also motivated his particular project—the project of defining the concept of number and the number one. There are two cases in which Frege's project would, nonetheless, be unnecessary. Frege's work would be unnecessary had the truths of arithmetic already been shown to be analytic. And Frege's would be the wrong sort of project if the truths of arithmetic could be shown to be analytic without definitions of the numbers. Frege addresses these issues in the last two sections of his first chapter.

Sections 15–17

In these sections, unlike the earlier sections of Chapter 1, the possibility of taking some of the truths of arithmetic to be primitive is not at issue. For these sections concern the views of writers who appear to have addressed directly the issue of whether the truths of arithmetic are analytic. These sections seem to have been written simply to convince Frege's readers that the possibility that the truths of arithmetic are analytic, in Frege's sense, has not really been considered by other writers. In section 15 Frege argues that neither Leibniz nor Jevons, the writers who seem to have made claims that entail that the truths of arithmetic are analytic, has offered a convincing argument for taking arithmetic to be analytic. In sections 16 and 17 Frege considers and rejects an argument of Mill's against the claim that arithmetical truths are analytic. Like Leibniz and Jevons, Mill does not really seem to be considering whether the truths of arithmetic could be analytic in Frege's sense. The grounds for rejecting Mill's argument seem to be simply that Mill does not recognize how much can be learned from logic.

I have proposed a framework for reading chapter 1 of *Foundations* in the preceding pages. On my reading, Frege's first chapter is meant to provide a motivation for his project and, on my reading, there are significant gaps in this motivation. Both the mathematical and philosophical motivations offered depend on an understanding of primitiveness on which the simplest claims of arithmetic, for example, that $0 \neq 1$, may not be primitive truths and on which it might be reasonable to require that these truths be proved from primitive truths. But it is not obvious what this notion of primitiveness is or why it is important. I have argued that there is no obvious mathematical reason for regarding simple arithmetical truths as nonprimitive and that Frege's discussion of the mathematical motivation of his project does not provide one. It is also not clear that, in Frege's writings, the related notions of provability and definability are to be understood as they are in the writings of other mathematicians. In addition, there is no obvious philosophical reason for refusing to regard any truth of arithmetic as a primitive truth. I have tried to show, however, that most of the first chapter of *Foundations* can be read as an argument that the simplest truths of arithmetic cannot be primitive truths. And such a reading of this chapter provides the beginnings of an account of the notion of primitiveness. In particular, this notion is closely bound up not only with an understanding of provability and definability but also with two apparently traditional philosophical distinctions: analytic/synthetic and a priori/a posteriori. But Frege's analytic/synthetic and a priori/a posteriori distinctions are no more obviously traditional philosophical distinctions than his notions of provability and definability are obviously mathematical notions. In order to get a sense of the significance of Frege's notion of primitiveness, it will be important to investigate his understanding of these distinctions.

CHAPTER 2

Laws of Thought

F REGE's overall project is to define the numbers and the concept of
number and to use these definitions to prove the truths of arithmetic from primitive truths. At this point, I have argued that his overall
project does not have a convincing mathematical motivation and that
chapter 1 of *Foundations* consists largely of an argument that the simplest truths of arithmetic are not primitive truths. I have also suggested
that the argument that the simplest truths of arithmetic are not primitive is meant to provide a philosophical motivation for undertaking
Frege's project. I have yet to explain what this philosophical motivation
is, however, and I have yet to give a characterization of Frege's notion of
primitiveness. It is now time to investigate these issues.

I believe that a serious motivation for Frege's project can be found via
the investigion of his claim that his reason for wanting to prove the
truths of arithmetic from primitive truths is that these proofs will enable
us to determine whether arithmetic is analytic or synthetic, a priori or a
posteriori. If this is so, the interest in Frege's project depends largely on
the interest of his understanding of these distinctions. It will be important to look more closely at how Frege understands the classification of
truths generated by these distinctions and whence derives his interest
in the classification of arithmetic truths.

A side benefit of this work is that it will begin to shed some light on
Frege's notion of primitiveness. For, although chapter 1 of *Foundations*
does not actually contain a characterization of primitiveness, the argu-

ments that the simplest truths of arithmetic are not primitive do tell us something about what it is for a truth to be primitive. In particular, one of the hallmarks of primitiveness is that it will be obvious from an understanding of the content of a primitive truth whether it is analytic or synthetic, a priori or a posteriori. It is this feature of primitive truths which makes Frege think the proofs he is after will enable him to determine the appropriate classification of the truths of arithmetic. But the result of my investigation of Frege's analytic/synthetic and a priori/a posteriori distinctions will still fall short of what is required for a characterization of primitive truth. Frege's understanding of primitive truth depends also on his notions of definability and provability. These notions will not be examined in detail until Chapter 3 herein.

Although the terms 'analytic', 'synthetic', 'a priori', and 'a posteriori' are traditional philosophical terms, it is not clear to what extent Frege's understanding of these terms belongs to any philosophical tradition. In fact, Frege explicitly mentions the divergence of his formulation of analytic/synthetic distinction from the Kantian distinction. But, given this self-conscious divergence, why should Frege have continued to use the traditional terms? To answer this question, it may help to see what evidence is available for determining how Frege viewed his work as fitting into a philosophical tradition.

Although Frege does not explicitly discuss the issue of how his project is related to traditional philosophical projects, his writings provide a number of clues. He says, in his introduction, that empiricists will like his views least (FA xi). He accords Mill, his representative empiricist, little respect and clearly means his arguments against Mill to be devastating. In contrast, although Frege argues against Kant's view as well as Mill's, his treatment of Kant is deferential. He claims in the first chapter of *Foundations* (FA 3) that he is only making clear what Kant, in particular, meant by 'analytic' and 'synthetic'. Near the conclusion of *Foundations* (FA 101) he says that Kant did a great service in drawing the analytic/synthetic distinction, but that Kant drew it too narrowly (FA 99). Indeed, Frege's arguments in *Foundations* against Kant's view that the truths of arithmetic are synthetic a priori are not meant to be devastating. He actually comes around to the view that these truths are synthetic a priori after his own project fails (see PW 279). Thus, there seems to be evidence that Frege viewed his own philosophical perspective as in some sense a Kantian perspective, and that he meant his arguments against Kant to be understood as taking care of some of the minor details on which Kant went wrong. "I have no wish to incur the reproach of picking petty quarrels with a genius to whom we must all look up with grateful awe" (FA 101). Of course, it is not clear how seriously this should be taken. Frege often seems to cite works of others for purely polemical purposes, and he was writing during a strong neo-Kantian movement in Germany.[1] It may not be unreasonable to suppose that Frege's sporadic nods to Kant are hooks designed to catch an audience of neo-Kantians.

I think that Frege's nods to Kant are more than strategic. But I will not be trying to sustain this judgement through a comparison of Kant's and Frege's epistemologies. Such a project is far beyond the scope of a chapter—and perhaps even that of a reasonably sized book—requiring, as it would, a reading of the epistemological doctrines of both writers.

Nor will I be trying to establish Frege's interpretation of Kant. There is simply insufficient evidence in Frege's writings on which to attribute to him any full-fledged interpretation of Kantian epistemology. Frege discusses only a few passages from Kant and these have a very narrow subject matter. Furthermore, while Frege's own philosophical picture is, in many ways, very grand, it does not possess the sort of detail one would expect to find in a doctrine that meant to provide an interpretation of even a modest fragment of Kant. It is possible, of course, that the fragmentary nature of Frege's discussion of Kant is symptomatic of his presupposing some interpretation—perhaps one that was in the air at the time he wrote. But, while this is entirely plausible, and while a survey of the writings of his contemporaries would no doubt provide likely interpretations, there is, again, insufficient evidence available to warrant the conclusion that Frege is assuming any particular interpretation. Thus, my argument for the sincerity of Frege's verbal bows to Kant rests on Frege's words alone—on the passages in his writings in which he discusses Kant.

The above qualifications may seem to make a discussion of Frege's nods to Kant peripheral to the issue of how Frege's own works are to be interpreted. After all, if there is no interesting relation to be drawn between Frege's views and Kant's, why is it of interest that Frege thinks there is? The interest I find lies in the nature of the views that Frege

[1]Hans Sluga gives a historical account of this movement in chapter 2 of *Gottlob Frege*. Sluga also argues that Frege's views were influenced by Kant and were, in fact, similar to those of some of his neo-Kantian contemporaries. The view that Frege's overall project was largely motivated by epistemological and, to some extent, Kantian concerns can also be found in Philip Kitcher, "Frege's Epistemology"; Tyler Burge, "Sinning against Frege"; Gregory Currie, *Frege: An Introduction to His Philosophy*.

takes to be his Kantian legacy. My point is not that Frege's philosophical motivation can be made clearer by examining a legacy of Kantian (or neo-Kantian) epistemology but, rather, that it can be made clearer by examining the passages in which Frege mentions Kant. I will try to show that an appeal to the evidence provided by these passages will make it possible to explain how Frege views the analytic/synthetic and a priori/a posteriori distinctions. This explanation will depend on our taking Frege's nods to Kant seriously. Not coincidentally, these are the passages in which the beginnings of a convincing philosophical motivation for Frege's overall project can be found. I will try to argue that Frege's interest in classifying arithmetical truth according to these distinctions is part of his interest in making an epistemological picture work. This epistemological picture emerges from Frege's attempt to fit together a series of passages from Kant.

Frege begins his discussion of the analytic/synthetic and a priori/a posteriori distinctions with the following words.

> It not uncommonly happens that we first discover the content of a proposition, and only later give the rigorous proof of it, on other and more difficult lines; and often this same proof also reveals more precisely the conditions restricting the validity of the original proposition. In general, therefore, the question of how we arrive at the content of a judgement should be kept distinct from the other question, Whence do we derive the justification for its assertion? (FA 3)

After distinguishing these questions, Frege continues: "Now these distinctions between a priori and a posteriori, synthetic and analytic, concern, as I see it, not the content of the judgement but the justification for making the judgement" (FA 3). And he adds, in a footnote, that he does not mean by this to be assigning a new sense to these terms but only to be stating accurately what Kant, in particular, meant by them.

Is this footnote disingenuous? It may seem that the divergence of Frege's formulation of the analytic/synthetic from Kant's is part of an attack on Kant.[2] In particular, Kant's formulation of the analytic/synthetic distinction seems to concern not the justification, but the content, of propositions. For a proposition is understood to be analytic, for Kant, when the predicate concept is contained in the subject concept. But this is not enough to show that Frege's formulation is meant to be part of an attack on Kant.

---

[2]Paul Benacerraf argues this in "Frege: The Last Logicist" (LL). All further references to this work in this chapter will be made parenthetically in the text.

Frege seems to think that the Kantian distinction *does* concern justification. In section 12 of *Foundations*, Frege understands Kant's claim that the truths of arithmetic are synthetic a priori as amounting to the claim that pure intuition must be invoked as "the ultimate ground of our knowledge of such judgements" (FA 18). In his introduction of his own distinction, he says that a judgement that a proposition is analytic or a posteriori is "a judgement about the ultimate ground upon which rests the justification for holding it to be true" (FA 3). Thus, Kant's views, as Frege understands them, seem fully to accord with Frege's characterization of his own claims about the classification of propositions. Of course, Frege's interpretation of Kant's claim in section 12 might itself be disingenuous—he might not *really* have believed Kant was concerned with justification. Frege might, after all, have believed this interpretation of Kant to be both inaccurate to Kant's actual views and useful for Frege's own purposes. But, as long as Frege consistently interpreted Kant's views in this way, such speculation is unproductive. It is not at all clear that it is important whether or not Frege is disingenuous in this sense. If Frege thought privately, but never publicly acknowledged, that the understanding of Kant which appears in his writings was inaccurate, this has only biographical interest. To show that Frege wants to fit together certain passages from Kant's work and use the result to motivate his project is to show that there is an interesting philosophical sense in which Frege viewed himself as a Kantian.

Not only would it be inaccurate to say that the Kantian analytic/synthetic distinction, as Frege understands it, does not concern justification, it would also be inaccurate to say that Frege's own distinction does not concern content. For it would be absurd to suppose that the content of a proposition is irrelevant to its ultimate justification.[3] In his actual investigation of the status of the truths of arithmetic, Frege discusses not only proofs but also the (often nonmathematical) grounds

[3]Of course, one might argue that the content of the terms 'black' and 'cat' is irrelevant to the justification of the claim that all black cats are cats. On the other hand, the content of these terms seems every bit as irrelevant to the determination that the predicate concept is contained in the subject concept. Furthermore, on Frege's view, some of the content of this claim does come into its justification. This content has to do with its logical structure. This case might look too easy because the subject concept is actually described by the use of two terms, one of which is used for the predicate concept as well. I will argue in later chapters that, given Frege's requirements for determining analyticity, it is necessary to define all expressions from primitive terms. Consequently, the process of determining whether, for instance, the claim that all bachelors are unmarried is analytic would require the definition of both 'bachelor' and 'unmarried' from primitive terms, and the result of using only primitive terms to restate the claim would make this claim as obvious as the claim that all black cats are cats.

that might be involved in the justification of the truths on which these proofs are based. The content of the proposition in question does come into this.

How, then, are we to understand Frege's claim that these distinctions concern justification and not content? It is important to note that Frege explicitly distinguishes, not content and justification, but the questions of *how we arrive* (*wie wir zu . . . kommen*) at the content and *whence we derive the justification* for its assertion (*woher wir die Berechtigung . . . nehmen*). Frege's claim that the analytic/synthetic distinction concerns the justification of a proposition, not its content, appears only after he makes this distinction. There are, of course, two interpretations of Frege's talk of discovering (arriving at) the content of a proposition (judgement). Frege could be talking either about coming to understand the proposition (judgement) or about coming to believe that it is true. Frege then goes on to say:

> When a proposition is called a posteriori or analytic in my sense, this is not a judgement about the conditions, psychological and physical, which have made it possible to form the content of the proposition in our consciousness; nor is it a judgement about the way in which some other man has come, perhaps erroneously, to believe it true; rather, it is a judgement about the ultimate ground upon which rests the justification for holding it to be true. (FA 3)

The question of the ultimate ground of the justification, as Frege understands it, is to be distinguished *both* from the question of how we come to understand some proposition *and* from the question of how we come to believe that it is true. It in no way follows, however, that the actual content of the proposition is not a factor in determining whether or not it is analytic. I will argue later in this chapter that there is a sense in which to say that a proposition is analytic is, for Frege, to say that it is justified by its content.

Thus, while Frege's formulation of the analytic/synthetic distinction appears to shift the emphasis from the content of a proposition to its justification, it is not clear that this is to be viewed as an attack on Kant. But if Frege is not meaning to attack Kant, it is important to figure out what motivates his reformulation of the Kantian distinction.

Frege's discussions of whether the truths of arithmetic are analytic, synthetic a priori, or synthetic a posteriori, all turn on what the ultimate ground of our justification of the truths of arithmetic could be. For instance, in order to convince his readers that the truths of arithmetic are not primitive and synthetic a priori, Frege argues that Kant has not shown convincingly that inner intuition can be used to justify the truths of arithmetic. The assumption is that if the truths of arithmetic are synthetic a priori, then inner intuition must be invoked in their justification. Similarly, in order to convince his readers that the truths of arithmetic are not synthetic a posteriori, Frege argues that the truths of arithmetic cannot be empirically justified. Finally, Frege says that if the truths of arithmetic are analytic (and, of course, he suggests that they are), this is to say that the ultimate ground of their justification is deeper than that of either the empirical sciences or geometry (FA 21). It is not clear what this ultimate ground is. Frege only suggests that it seems that the laws of number must be intimately connected with the laws of thought. I will use the term 'thought' for the ultimate ground of analytic truths. Thus, it seems that, for Frege, the analytic/synthetic distinction, in combination with the a priori/a posteriori distinction, allows us to categorize propositions according to the sort of ultimate ground that must be invoked in order to justify them. Since there are no analytic a posteriori propositions, it follows that Frege is assuming that there can be no more than three sorts of ultimate ground.

The three sorts of ultimate ground, then, are sense perceptions (for propositions that are synthetic a posteriori); inner intuition (for propositions that are synthetic a priori); and thought (for propositions that are analytic). But these do not constitute three independent realms. Frege's discussion in section 14 suggests that to identify a proposition as synthetic is not to say that analytic laws—the most general laws of thought—do not apply to it. These laws govern everything thinkable and hence are applicable to all assertions. Similarly, a synthetic proposition that is justified a posteriori is still subject to the laws of geometry— laws whose ultimate ground is inner intuition. Indeed, both laws of thought and laws of geometry constitute parts of an a posteriori justification. Thus, the division of ultimate grounds constitutes a hierarchy of generality, and the classification of a truth depends on how far down one must go in this hierarchy in order to justify it.[4] Synthetic truths, for instance, require for their justification laws that are less generally applicable than the laws that apply to all thought. But this does not tell us how the distinct sorts of ultimate ground are to be characterized or how are they invoked in our justification. In order to fill in these details, it is necessary to look at what is involved in justification.

Since justification, for mathematics, consists in proof (see FA 4), it

---

[4]This is also suggested by Frege's talk of the intuitive idea of a sequence in section 23 of *Begriffsschrift*.

may be reasonable to begin by asking about how an ultimate ground is invoked in a proof. Frege says, in section 3 of *Foundations*, that the problem is to find the proof of a proposition and follow it back to the primitive truths. The ultimate ground of the proposition will depend on the nature of these primitive truths. But all this work may seem unnecessary. Many mathematical proofs include a step that is explicitly licensed by "intuition." Does this show that intuition is included in the ultimate ground?

Frege believes that although mathematical proofs sometimes appear to include appeals to inner intuition, these appeals are misleading. In particular, while many mathematicians include steps for which the only explicit justification offered is "intuition," such appeals to intuition usually simply indicate that the mathematician believes the transition to the next step is obvious. Frege says, of some transitions in proofs,

> Often, nevertheless, the correctness of such a transition is immediately self-evident to us, without our ever becoming conscious of the subordinate steps condensed within it; whereupon, since it does not obviously conform to any of the recognized types of logical inference, we are prepared to accept its self-evidence forthwith as intuitive, and the conclusion itself as a synthetic truth—and this even when obviously it holds good of much more than merely what can be intuited. (FA 102–103)

In these cases there may be no actual content to the apparent invocation of intuition. A mathematician usually means by "we can see by intuition that" nothing more than "it is obvious that." But it does not follow that intuition is not necessary for the transition. Part of Frege's project is to make it clear, in such cases, whether the appeal to intuition is simply an indication that the transition to the next step requires a series of simple steps omitted for purposes of clarity or whether a substantive presupposition is necessary. Frege's *Begriffsschrift* is a preliminary step toward addressing such issues. This part of Frege's project, he says in the preface to *Begriffsschrift*, led to his developing his logical notation, for

> to prevent anything intuitive from penetrating here unnoticed, I had to bend every effort to keep the chain of inferences free of gaps. In attempting to comply with this requirement in the strictest possible way I found the inadequacy of language to be an obstacle; no matter how unwieldy the expressions I was ready to accept, I was less and less able, as the relations became more and more complex, to attain the precision that my purpose required. This deficiency led me to the idea of the present ideography. Its

first purpose, therefore, is to provide us with the most reliable test of the validity of a chain of inferences and to point out every presupposition that tried to sneak in unnoticed, so that its origin can be investigated. (BEG 5–6)

To identify a proof that requires an appeal to intuition would, it seems, amount to finding, in the proof, a hitherto unstated presupposition whose ultimate ground is intuition. Intuition is not, on Frege's view, a special sort of reasoning. He says, "Thought is in essentials the same everywhere: it is not true that there are different kinds of laws of thought to suit the different kinds of objects thought about" (FA iii). The identification of intuition as part of the ultimate ground of some mathematical truth will not depend on an explicit appeal to intuition or intuitive reasoning in the proof but, rather, on the necessity of a presupposition that has intuition as its ultimate ground. But this does not answer the question about how intuition might be invoked, it only pushes it back.

How is the origin of a presupposition to be investigated and when will such investigation allow us to conclude that the presupposition has intuition as its ultimate ground? The first step is to prove the presupposition if it is provable—to follow it back to the primitive truths on which *it* depends. But it is important to note that the ground of a truth does not consist of the truths on which it depends and the ultimate ground of a truth does not consist of the primitive truths on which it depends. It is not that, on Frege's view, a primitive truth is its own ground but, rather, that an understanding of a primitive truth will make it clear what sort of ground that truth requires for its justification. Our proofs stop when the primitive truths are reached. How, then, are we to talk about the justification of these truths? Frege does not discuss this in *Foundations*. He does, however, mention it in an unpublished work. He says:

> Now the grounds which justify the recognition of a truth often reside in other truths which have already been recognized. But if there are any truths recognized by us at all, this cannot be the only form that justification takes. There must be judgements whose justification rests on something else, if they stand in need of justification at all. (PW 3)

But what could this "something else" be? Where could we find an account of the something else that is the ground of the noninferential justification of truths not justified by other truths? Frege continues.

And this is where epistemology comes in. (PW 3)

Epistemology, then, is the field to which the investigation of ultimate grounds and the identification of the ultimate grounds of the unprovable truths belongs. In order to determine what is the ultimate ground of the truths of arithmetic, as Frege intends, it is necessary to prove these truths from primitive truths and then to identify the ultimate grounds of these primitive truths. The latter task belongs to epistemology. But Frege rarely discusses actual unprovable truths and does not expend much effort in the attempt to establish the ultimate ground of any unprovable truth. Although one of the hallmarks of a primitive truth, for Frege, is that the category into which its ultimate ground falls is obvious, Frege does not tell us what makes this category obvious. It seems, then, that Frege is missing an account of the sorts of available ultimate ground and how an ultimate ground can justify a primitive truth. Without such an account, how could the category of the ultimate ground of a primitive truth be evident? The point at which epistemology, on Frege's view, comes in to an investigation of justification is precisely the point at which, in his own discussions of justification, Frege is mute.

How are we to regard this fundamental gap in Frege's motivation? There seem to be two options. Either one could infer that Frege's motivation stops here—that his interests are confined to inferential justification and hence are not truly epistemological—or one could infer that Frege thought someone else had addressed the issue of the ultimate grounds of the justification of primitive truths. To take the first option, of course, is to leave a number of questions unanswered. If Frege truly has no views about the ultimate grounds of primitive truths, he has no reason to think his investigation of the primitive truths on which arithmetic depends will settle the question of the ultimate ground of the truths of arithmetic. More significant, there will be no reason to deny that the simplest truths of arithmetic are primitive. Thus, the first chapter of *Foundations* has no obvious role to play, and Frege's attempt to precede his introduction of definitions of the numbers with a discussion motivating his search for these definitions must fail. Consequently, the more attractive option is the second. If Frege's work can be viewed as a contribution to an epistemological picture that he believed was already available, there might be no need for him to discuss the epistemological status of primitive truths. The antecedent epistemological picture would allow him to view his proofs of the truths of arithmetic from primitive truths as the final step in determining their ultimate ground.

If Frege viewed his work as a contribution to an epistemological picture, the evidence from his comments about the writings of others suggests that Frege would have taken this picture to be a Kantian picture. This is the reason for my concern to argue that Frege did not view his reformulation of the analytic/synthetic distinction as an attack on Kant. In the remainder of this chapter I will examine the evidence for reading Frege's work as an attempt to contribute to a picture that he viewed as Kantian, and I will argue that such a reading of Frege's work provides an explanation for most of the questions Frege asked and his strategies for answering them.

In a 1979 paper, "Frege's Epistemology,"[5] Philip Kitcher concludes, from an examination of Frege's motivation, that Frege must have viewed his work as a contribution to a Kantian epistemological picture. Kitcher's arguments for his conclusion are similar to some of the arguments I have given above. He argues that Frege's project should be seen as being epistemologically motivated; on the other hand, he also claims that Frege does not discuss the basic issues of epistemology which are intimately connected to his project. Kitcher says that these issues are how we can come to know the laws of logic and how our knowledge of the laws of logic can be invulnerable to doubt (FE 240–241). Kitcher suggests that if Frege's interest is epistemological it seems very odd that he should avoid these basic epistemological issues. Kitcher's solution to this puzzle is to understand Frege as contributing to a philosophical picture in which all the basic epistemological questions had been answered by Kant. The similarity of this argument, at least in outline, to the argument I have given above is significant. I agree with Kitcher about the role that some view of Kantian epistemology plays in Frege's picture. Nonetheless, the epistemological picture Kitcher ascribes to Frege is very different from the picture I want to ascribe to Frege. In addition, I think Frege does address the issues Kitcher claims he does not discuss. In order to make my own view clearer, it will be useful to discuss Kitcher's.

Kitcher describes the Kantian analytic/synthetic distinction, in combination with the a priori/a posteriori distinction, as generating a three-fold "epistemological division among ways in which propositions can

[5]"Frege's Epistemology," *Philosophical Review*, 88 (1979) (FE). All further references to this work in this chapter will be made parenthetically in the text.

be known" (FE 250). Kitcher describes this division as a division of propositions according to the source of our knowledge of them. And Kitcher attributes to Frege a desire to categorize the truths of arithmetic according to this division.

Although my discussion of Frege's attempt to find the ultimate ground of the truths of arithmetic depends on a similar division, I have not, thus far, used the expression "source of knowledge" to describe this division. This is because Frege rarely talks about knowledge in *Foundations*. Although he occasionally uses the term *Erkenntnisgrund*, which can be translated as 'ground of knowledge' (see, e.g., FA 18), most of his talk about grounds concerns justification (*Berechtigung*), which, for purposes of mathematics, amounts to proof. Frege does not talk about sources of knowledge (*Erkenntnisquellen*) in *Foundations*. This is not to say that Frege avoids this expression altogether. In his correspondence, Frege uses this expression at least twice in characterizations of his project (COR 55/WB 87, COR 57/WB 89). He also uses it in a letter to Hilbert about their different views of the axioms of geometry (COR 37/WB 63). Furthermore, Frege continued to write about the foundations of arithmetic after he was convinced of the failure of his project to show that the truths of arithmetic could be proved from logical laws and definitions alone. In two of his later, unpublished, discussions of the foundations of arithmetic, he begins by distinguishing three sources of knowledge: sense perception, the logical source, and the geometrical source.[6] These might be taken to correspond to the three sorts of ultimate ground discussed above. Moreover, they appear in the context of discussions of whether appeals to sense perception or intuition are needed in proofs of arithmetic. Thus, it is not at all clear that there is any important difference between this later use of "source of knowledge" and Frege's earlier use of 'ultimate ground'. In later chapters I will be using the expressions interchangeably. However, there is a problem with the way Kitcher interprets "source of knowledge."

Kitcher interprets talk about the sources of knowledge as talk about the process by which one comes to judge that something is true. His reason for interpreting sources of knowledge in this way seems to derive from his conception of epistemology and, in particular, of the connection of epistemology with knowledge. On Kitcher's view, epistemology concerns the mental states or processes that produce knowledge. He attributes a "psychologistic account of knowledge" (FE 243) to

6See PW 267–274/NS 286–294, PW 278–281/NS 298–302.

Frege from which it follows that "the question of whether a person's true belief counts as knowledge depends on whether or not the presence of that true belief can be explained in an appropriate fashion" (FE 243). Kitcher holds that a central task for any theory of mathematical knowledge is that of giving "a precise characterization of the special states and/or processes" that give rise to mathematical knowledge (FE 244). On this view, the appropriate states and/or processes constitute the justification of a truth. And the written proof of a mathematical truth is simply a public record of a certain sort of psychological process. The purpose of *Begriffsschrift*, on Kitcher's view, is to provide the laws governing the correct psychological processes to be employed in inference. Frege's concern, according to Kitcher, is to induce mathematicians to use correct psychological processes in their proofs and to show that a certain sort of psychological process will suffice to justify the truths of arithmetic.

But the evidence for tying Frege's analytic/synthetic distinction to the status of the process that produces a person's true beliefs is very slim. Frege rarely speaks of the process by which a person comes to believe some truth except to say that this is a separate issue from the justification of that truth. Kitcher's explanation of these passages is that their purpose is not to distinguish the process by which one comes to believe a truth from its justification (where 'justification' has some nonpsychological sense) but, rather, to distinguish the process by which one in fact comes to believe it from the process by which one *ought* to come to believe it. The point is that Frege is concerned with the laws of thought not in a descriptive, but in a normative, sense.

In support of his interpretation, Kitcher calls attention to Frege's remarks that inference is a mental process. Since the normative laws are to provide rules for correct inference, it sounds as if the rules of inference are supposed to be rules governing how this mental process ought to be carried out. But, while it is not particularly controversial to say that our making inferences is a psychological process, this does not imply that an interest in correct inference is an interest in this psychological process. Frege's logical laws contain no references to beliefs or mental process. Indeed, no discussion of mental processes is required in order to determine whether a proof is the real justification of some truth. Kitcher's claim (FE 242) that written proofs correspond to special kinds of psychological processes may be correct but, even if it is, Frege seems not to be concerned with the psychological processes at all. He says, "But above all we should be wary of the view that it is the business of logic to investigate how we actually think and judge when we are in agreement with the laws of truth" (PW 146). Frege goes on to suggest that if we were concerned with these psychological processes that are in agreement with the laws of truth, we would have to go back and forth between logical laws and these psychological processes. Frege does not engage in this activity. Furthermore, this is not an indication that he believes this activity has already been adequately carried out by someone else. In fact, Frege suggests that to do this would result in our being "seduced into asking questions with no clear meaning" (PW 146).[7]

But if the laws of logic are not laws governing the mental processes we should employ in our inferences, how are we to understand Frege's characterization of the laws of logic as normative? As Kitcher notes, Frege's primary concern in these discussions is to avoid having his laws

[7]Given the enigmatic nature of this claim, it is probably important for me to note that my only purpose in quoting it is to demonstrate the evidence that Frege did not regard the enterprise of comparing logical laws and psychological processes as a part of logic. I have not given an explanation of Frege's claim, however. For, although it seems clear that Frege does not mean to endorse such a project, it is not clear exactly what his objection is. In fact, Frege says very little about this. The entire paragraph is: "But above all we should be wary of the view that it is the business of logic to investigate how we actually think and judge when we are in agreement with the laws of truth. If that were so, we should have constantly to have one eye on the one thing and one eye on the other, and continue paying attention to the latter whilst taking a sidelong glance at the former, and in the process we should easily lose sight of a definite goal altogether. We should be seduced into asking questions with no clear meaning and as a result a satisfactory outcome to our investigations would be as good as impossible" (PW 146). I think that there is insufficient evidence here to make out a substantive objection to attribute to Frege. I do, however, have a suggestion about what Frege might mean.

Suppose it is the business of logic to investigate how we actually think when we are in agreement with the laws of truth. Then, presumably, there is no independent science that is concerned with the laws of truth. Nonetheless, the laws of truth are, on Frege's view, distinct from mental processes. Frege's point, I suspect, is that we can only investigate how we think when we are in agreement with the laws of truth if we can, independently, establish both what the laws of truth are and how our actual thought processes work.

If we were forced to investigate both areas simultaneously—for instance, to use the identification of a correct inference for purposes both of identifying the mental processes involved in someone's making the inference and of identifying the general, nonmental, laws that might license the inference, we might be led to confuse these separate issues. The significance of this confusion is that the nonmental laws identified in such a procedure have a very different status from the mental processes identified. For there is no way to be sure, short of a survey of all correct inferences (or an independent summary of the laws of truth), that the mental process identified will always result in a correct inference. On the other hand, there *is* a way of determining whether application of the nonmental law will always result in a correct inference. As I argue later in this chapter, for Frege a law of truth can be identified by its content.

of thought interpreted as generalizations of how people actually think. The relationship between Frege's laws of thought and actual thinking is meant to be different from the relationship between the laws of nature and actual events. If a particular event violates what purports to be the statement of some law of nature, then the statement does not express a law of nature. On the other hand, if the inferences actually carried out by some mathematician violate one of Frege's laws of thought, this is, on Frege's view, no evidence of any defect in the expression of the law. The relationship between Frege's laws of thought and actual thinking is meant to be analogous to the relationship between ethical laws and actual actions. Frege says:

> Logic has a closer affinity with ethics. . . . Although our actions and endeavours are all causally conditioned and explicable in psychological terms, they do not all deserve to be called good. Here, too, we can talk of justification and here, too, this is not simply a matter of relating what actually took place or of showing that things had to happen as they did and not in any other way. . . . What makes us so prone to embrace such erroneous views is that we define the task of logic as the investigation of thought, whilst understanding by this expression something on the same footing as the laws of nature: we understand them as laws in accordance with which thinking actually takes place. (PW 4)

Frege repeats this sort of remark throughout his career. An example is this later comment about the laws of logic:

> So if we call them laws of thought or, better, laws of judgement, we must not forget we are concerned here with the laws which, like the principles of morals or the laws of the state, prescribe how we are to act, and do not, like the laws of nature, define the actual course of events. (PW 145)

The study of logic is to be quite independent of any facts about the psychological processes by which thinking actually takes place. All of this, of course, fits with Kitcher's interpretation. For Kitcher attributes to Frege an interest not in the actual psychological processes by which mathematicians make their inferences but, rather, in the psychological processes that mathematicians *should* use in their inferences—the psychological processes that actually justify mathematical truths.

But it is also important to realize that the comparison with ethics is not meant to prohibit any comparison between Frege's laws of thought and natural laws. In particular, Frege does not mean to suggest that to say that the laws of logic are normative is to distinguish them from other scientific laws. For he also says, "We could, with equal justice, think of the laws of geometry and the laws of physics as laws of thought or laws of judgement" (PW 145). Laws of geometry and physics are not laws about the psychological processes to be used in inferences concerning geometry and physics. These laws are descriptive. And the laws of logic are no less descriptive than the laws of physics. Just as any natural event that violates the apparent statement of a natural law exhibits a defect in the statement, similarly any *correct* inference that violates a statement of a logical law (or Frege's one rule of inference, *modus ponens*) would exhibit a defect in that statement. Frege's laws of thought are meant to be generalizations—not generalizations of actual processes of judging something to be true but generalizations of truths. The truths themselves are distinct from the process of judging. His rule of inference is meant to be a generalization of correct inference, that is, legitimate transition from one truth to another. Once again, the generalization is over the truths, not the actual thought processes involved in the activity of inferring. Thus, in his later writings, Frege calls these laws "laws of truth" (see, e.g., CP 351). He says:

> Any law asserting what is, can be conceived as prescribing that one ought to think in conformity with it, and is thus in that sense a law of thought. This holds for laws of geometry and physics no less than for laws of logic. The latter have a special title to the name "laws of thought" only if we mean to assert that they are the most general laws, which prescribe universally the way in which one ought to think if one is to think at all. (BLA 12)

Although we ought to judge in accord with these laws of truth, these laws do not straightforwardly address our mental processes any more than the laws of physics address our mental processes.

To see this, it may help to consider an example. Frege's laws of thought are meant (along with his rule of inference) to license the transition from a universal generalization to an instance of the generalization.[8] From the claim that everything is either red or not-red we are allowed to infer that Frege is either red or not-red. But the details of the

---

[8]Actually, this is oversimplified. Frege's laws, along with his rule of inference, *modus ponens*, will not suffice to license such a transition. In fact, in order to use Frege's Begriffsschrift to demonstrate that the transition is legitimate, it would also be necessary to use his rules for the use of the signs of Begriffsschrift. Since all of these rules are, in effect, fundamental principles of thought, their use has the same status as the use of a law of thought.

psychological process involved when this inference is carried out are unimportant. The significant features of this inference are exhausted by the contents of the two judgements. There is no need, in order to determine whether a truth is justified by an inference, to discuss the mental states of a person who makes this inference. And there is no role for such a discussion to play—these mental states are irrelevant.

Thus, it does not follow that, because inference is a mental process and Frege's analytic/synthetic distinction concerns inferential justification, this distinction concerns the mental process by which a belief is either produced or justified. Kitcher's claim that Frege was concerned with such mental processes cannot be supported by Frege's discussion of inferential justification. But this does not show that Kitcher is entirely wrong.

It is important to remember that inference alone is insufficient for justification. An inference licenses its conclusion only if the premises of the inference are truths.[9] Although Frege's explicit discussions of justification almost invariably concern inference, not all justification, on Frege's view, is inferential. What I have said so far does not yet clearly preclude the possibility that, once we have inferred some truth from primitive truths, the ultimate grounds of the justification of *these* truths might turn on psychological processes. Frege says very little about the ultimate grounds of primitive truths. Thus, Frege's apparent lack of interest in mental processes may simply be a result of his focus on inferential justification.

Furthermore, there is some indication in Frege's writings that he thinks some truths are justified via a process of mental picturing. In his discussion in section 5 of *Foundations* of whether the truths of arithmetic could be synthetic a priori, he seems to be talking about whether the constructions of mental pictures of arrangements of fingers or points can, without proof, justify numerical formulae. Thus, even if some of the details of Kitcher's interpretation are wrong, there looks to be some support for Kitcher's attribution to Frege of a psychologistic epistemology.

I think, however, that these passages do not support such an attribution. Later in this chapter I will argue that Frege is committed to a nonpsychologistic view about the ultimate ground for the justification

[9]Although this is true, it might seem to suggest that there are inferences whose premises are false. On Frege's view, however, an inference must have true premises. This is not to say that we cannot engage in reductio proofs but, rather, that reductio proofs do not have falsehoods as premises. See, for instance, CP 375.

of primitive logical laws and that it follows that Frege does not abandon, for primitive truths, his distinction between the ultimate ground of their justification and the process by which we may (or, even, must) come to believe them. In order to understand Frege's views on the ultimate ground of the primitive logical laws, it will be necessary to examine more closely the aims and pretensions of Frege's *Begriffsschrift*. I suggested earlier that a productive strategy for this sort of examination is to begin with a look at his remarks about Kant, and I will adopt this strategy now. The adoption of this strategy will provide an additional argument against Kitcher's view. For one of Kitcher's central arguments for the attribution of a psychologistic epistemology to Frege has to do with an account of the relation between Frege and Kant—an account that, I will argue, is not substantiated by Frege's texts.[10]

As I have already indicated, I agree with Kitcher that Frege's general lack of attention to primitive truths is explicable if we assume that Frege thought Kant had answered the questions about the ultimate ground of primitive synthetic a priori truths. In his discussion of the relation between Frege's views and Kant's, Kitcher acknowledges the existence of Frege's many vehement arguments against psychologism but suggests that Kantian epistemology is psychologistic (FE 243). This seems to create a difficulty for the claim that, on Frege's view, Kant had answered the basic epistemological questions. Kitcher's solution is to suggest that the scope of Frege's antipsychologism was sufficiently restricted to allow Frege a Kantian (psychologistic) view of knowledge. I will argue that Kitcher is wrong about the restrictions to Frege's antipsychologism and, furthermore, that Frege construed his antipsychologism as a Kantian view.

The first thing to note is that, in his descriptions both of logic and of the analytic/synthetic distinction, Frege's concentration on the ultimate justification of a truth rather than on the psychological processes by which we in fact convince ourselves of it seems entirely consonant with Frege's claims to be following Kant. Frege refers to Kant's *Logic* in

---

[10]In fact, Kitcher believes that Kantian epistemology is psychologistic and that Frege has adopted Kant's answers to basic epistemological questions. Thus, Kitcher must explain Frege's apparent antipsychologism. Kitcher's solution, as I discuss below, is to argue that Frege's antipsychologism was restricted in scope. Benacerraf also suggests that Kantian epistemology is psychologistic (LL 26). Unlike Kitcher, however, Benacerraf construes Frege's antipsychologism as anti-Kantianism. As I have indicated earlier, questions about Kant's actual view—whether or not Kantian epistemology is really psychologistic, for instance—are outside the realm of my investigation. One of my aims is to argue that Kant, as Frege reads him, is every bit as antipsychologistic as Frege.

Foundations (FA 19), and his characterization of the laws of thought as concerning actual thinking in a normative rather than descriptive sense are virtually paraphrases of remarks like the following from Kant's *Logic*.

> Some logicians presuppose psychological principles in logic. But to bring such principles into logic is as absurd as taking morality from life. If we took the principles from psychology, i.e., from observations about our understanding, we would merely see how thinking occurs and how it is under manifold hindrances and conditions. . . . In logic, the question is not one of contingent but of necessary rules, not how we think, but how we ought to think.

And

> In logic we do not want to know how the understanding is and thinks and how it hitherto has proceeded in thinking, but how it ought to proceed in thinking. Logic shall teach us the right use of the understanding, i.e., the one that agrees with itself. (*Logic* 15)

These statements, of course, do not conflict with Kitcher's interpretation of Kantian epistemology as psychologistic because they do not suggest explicitly, as Frege's statements do, that the laws of logic are also descriptive of something nonmental. However, although these statements may be consistent with Kitcher's interpretation of Kantian epistemology as psychologistic, they are also consistent with Frege's statements that are meant to fit in with a nonpsychologistic account of inference. For it is clear, from the passages quoted in the last few pages, that Frege thought the issue of how we ought to think, or the right use of the understanding, could be addressed by his nonpsychologistic logic.

In order to address the issue of how, if not via mental processes, the primitive logical truths can be justified, it will be important to look more closely at Frege's reason for diverging, in *Begriffsschrift*, from the traditional Aristotelian logic and at the consequences of this divergence for Frege's view of Kantian epistemology.

Frege says in the preface to *Begriffsschrift* that his development of this new logic arose from his desire to determine whether arithmetic was analytic or synthetic. But the complex view of the ultimate justification of analytic truths which emerges from Frege's *Begriffsschrift* presents one of the greatest difficulties in connecting Frege's threefold division of the ultimate grounds of analytic, synthetic a priori, and synthetic a posteriori truths with Kant. This difficulty is heightened if we refer, as Frege did in some of his later unpublished writings, to a logical source of knowledge. For it is not at all clear that, on the traditional Kantian distinction, there is any room for a serious justification of analytic truths. Frege was aware of this. He alludes (FA 101) to a passage of the *Critique of Pure Reason* (A 8/B 12) in which Kant seems to indicate that our knowledge can be extended only by synthetic truths. But Frege would count the conclusions of purely logical inferences as analytic truths, and these conclusions, Frege says, do extend our knowledge. How, then, can Frege have written (FA 3) that he means to be stating accurately what Kant meant by 'analytic'?

This issue is addressed, briefly, in section 88 of *Foundations*. Frege says there that Kant had an inkling of the wider sense in which Frege uses the term. The evidence is that "[B 14] he says that a synthetic proposition can only be seen to be true by the law of contradiction, if another synthetic proposition is presupposed" (FA 100). In the next few paragraphs of *Foundations* Frege suggests that Kant's remarks should be interpreted as indicating that Kant believed our knowledge of analytic truths can be derived from logical laws and definitions of the concepts involved alone, whereas, in the case of synthetic truths, some additional support or evidence is needed. This might stem from an interpretation of the following argument Kant offers in support of the claim that arithmetic is synthetic: "For it is obvious that, however we might turn and twist our concepts, we could never, by the mere analysis of them, and without the aid of intuition, discover what (the number is that) is the sum" (B 16). But if Kant thought, as Frege seems to suggest, that the conclusions of arguments that rely only on logical laws and definitions are analytic, why does Kant's notion of analyticity differ from Frege's?

The obvious answer is that Frege's laws of logic license more inferences than the Aristotelian logic that Kant assumed was correct. But this difference in logics might suggest that Frege's notion of analyticity is, in fact, not importantly related to Kant's at all. In order to see how Frege might have viewed himself as saying what Kant really meant by 'analytic', it is important to look at some of the details of the difference between Frege's new logic and Aristotelian logic.

Frege says, in the preface to *Begriffsschrift*, that he found ordinary language to be inadequate for purposes of making clear precisely what presuppositions were necessary for the transitions in mathematical proofs. In Aristotelian logic, however, the grammatical regimentation of a sentence into subject and predicate was thought to exhibit all the structure necessary to determine all the correct purely logical inferences in which the sentence can be used. For purposes of evaluating inferences according to an Aristotelian logic, the expressive power of ordinary language is sufficient. One of the fundamental divergences of Frege's logic from Aristotelian logic is the introduction of a nongrammatical regimentation of sentences into function and argument expressions. This regimentation allowed Frege to exhibit the validity of inferences involving multiple generality—inferences that were becoming increasingly important in nineteenth-century mathematics. A consequence of Frege's new notation is that not all truths can be regarded, for logical purposes, as being constructed out of a subject and predicate.

Now it is easy to see why Frege might have viewed the introduction of his logic as causing difficulties for Kant. The analytic/synthetic distinction, as formulated by Kant in his *Critique of Pure Reason*, is a distinction between those judgements in which the predicate concept is contained in the subject concept and those in which the predicate concept is not contained in the subject concept (CPR A 6/B 10). If logic is Aristotelian logic, and if our knowledge of analytic truths can be derived from logical laws and definitions, then the only propositions that can be analytic are those that have a subject concept and a predicate concept. Such propositions can be expressed by sentences that are grammatically regimentable into subject and predicate, and this grammatical regimentation mirrors the logical regimentation. On the other hand, Frege's Begriffsschrift licenses inferences involving propositions that do not have subject and predicate concepts although they may be expressed by sentences that can be regimented, grammatically, into subject and predicate. Given this divergence between grammatical and logical structure, which is to determine analyticity?

Were Frege to continue to take grammar as a guide to analyticity, his new logic would not require a reformulation of the Kantian distinction. Why should he not adopt this strategy? One of the consequences of such a strategy would be that some truths that follow from Frege's logic alone using only definitions will not be analytic. For it is possible to prove, in Frege's logic and using only definitions, some truths whose logical regimentation cannot be expressed by reference to subject and predicate concepts. If the distinction is understood as just suggested, then at least some of the details of Frege's logic will have to be counted as additional support. Because of this, the reformulation suggested here will violate the spirit of Kant's distinction as understood by Frege. For, as Frege understands the distinction, its purpose is to draw the line between those propositions whose truth is determined by the content of the concepts involved—those truths that can be justified by "turning and twisting the concepts involved"—and those for which additional support is required. Part of Frege's aim is to show that our concepts are more highly structured than they had previously been taken to be and that, as a result, there is new knowledge to be gained by virtue of turning and twisting them. Frege is, in a sense, introducing a new source of knowledge: the logical source. On Frege's view, analyticity must be determined by logical structure.

Thus, Frege objects to Kant's formulation of the analytic/synthetic distinction (FA 100) because it is not exhaustive. The criteria that determine analyticity must, on Frege's view, be determined by logical, not grammatical, structure. Given this assumption, the Kantian formulation is exhaustive if an Aristotelian logic is the true logic, but it is not exhaustive if Frege's logic is the true logic. Since Frege is interpreting the analytic/synthetic distinction as a means for distinguishing between those truths we could have knowledge of on the basis of having an understanding of the concepts involved and those we could have knowledge of only through some additional support, it is important that this distinction be exhaustive. Frege's reformulation provides an exhaustive distinction. For a truth either is or is not provable from definitions and the laws of logic alone. It remains to argue, however, that Frege's reformulation of the Kantian distinction is designed to preserve what he takes to be the central features of the distinction. That is, it remains to argue that the epistemological role of logic precludes taking logic as additional support.

Frege begins the preface of *Begriffsschrift* by drawing the distinction between the justification of a proposition and the history of an individual's belief in it. He then introduces his project of determining whether the truths of arithmetic can be justified purely by means of logic— "those laws on which all knowledge rests" (BEG 5)—or whether they require facts of experience for their justification. In order to determine this, he says, "I first had to ascertain how far one could proceed in arithmetic by means of inferences alone, with the sole support of those laws of thought that transcend all particulars" (BEG 5). The project of *Begriffsschrift* was to introduce a new language for use in mathematics (and, eventually, all science) which avoided the imprecision of ordinary language, and to express, in this new language, the laws of thought which transcend all particulars—or, as Frege says later, "the most general laws of truth" (PW 128). Furthermore, this precise language was to be formulated in such a way that scientific claims could be expressed so clearly and unambiguously that it would always be possible to determine whether a chain of inferences follows without the need of any extra assumptions or presuppositions from these laws of thought. If the inference does follow from the laws of Frege's *Begriffsschrift*, this suffices to determine its justification. These laws are meant to be applicable to actual mathematical proofs.

For instance, in "On the Scientific Justification of a Conceptual Notation (Begriffsschrift)," a paper published after *Begriffsschrift* but before *Foundations*, Frege gives examples of unstated presuppositions that appear in proofs from Euclid's *The Elements*. Frege says,

> Only by paying particular attention, however, can the reader become aware of the omission of these sentences, especially since they seem so close to being as fundamental as the laws of thought that they are used just like those laws themselves. (CN 85)

Frege's claim is that, were Euclid's proofs expressed in Begriffsschrift, the omission would be obvious. In general, once an inference is expressed in Frege's logical language, either it can be seen to follow by these laws of thought or it will be evident that some hitherto unexpressed presupposition needs to be invoked in order to justify the inference. The application of the fundamental laws of thought in the inference does not amount to the recognition of unexpressed presuppositions or additional support. If the logical laws alone suffice, this shows that no additional support is needed.

Frege's view of the laws of logic as completely general also seems to fit in with some of the remarks in Kant's *Logic*. For instance: "Now this science of the necessary laws of understanding and reason in general, or—which is the same—of the mere form of thinking, we call *logic*" (*Logic* 15). And logic is referred to as a "science concerning all thinking in general, regardless of objects as the matter of thinking," and logic "abstracts entirely from all objects." How is Frege's view of the generality of logic to be understood?

Frege does not, of course, intend to set out a law for every generalization of a logical truth. Some logical laws can be derived from others, and Frege says that it would not be possible to list all the laws of thought (BEG 29). Rather, his aim is to set out a small number of laws from which the rest can be deduced. Frege's mention here of "all laws of thought" may seem to suggest that he is requiring that all thought (or at least all scientific thought) be reducible logic. But this is not quite right. It is not that all scientific conclusions must be justified by the laws of logic alone. Rather, all scientific conclusions must be justified by the laws of logic along with whatever facts of experience may be appropriate or necessary. The inferences themselves—the transition from some truths to other truths—however, must be justified by logic alone. And, as already noted, Frege denies that there are "different kinds of laws of thought to suit the different kinds of objects thought about" (FA iii). This may seem unreasonable. In particular, this may seem to rule out the use of induction in natural sciences as illegitimate. He certainly does seem to have no room for a notion of inductive inferences. On the other hand, the procedure of induction (*Verfahren der Induction*) is mentioned in *Foundations* (FA 16), and he talks of 'inductive truths'. How is this to be understood?

After distinguishing scientific induction from "a mere process of habituation," he says "Induction must base itself on the theory of probability, since it can never render a proposition more than probable" (FA 16). Presumably, Frege means by this that the apparent conclusion of an inductive argument cannot be detached. From the information that an urn contains 99999 black balls and 1 white ball, we are not entitled to conclude, however heavily we may be willing to bet on it, that it is true that a ball drawn from the urn by a random procedure will be black. But, while it seems harmless to prohibit this particular conclusion, Frege's view may seem to commit him to denying that the truth of a theory of natural science can be established by inductive support.

He says, in the preface to *Begriffsschrift*,

> In apprehending a scientific truth we pass, as a rule, through various degrees of certitude. Perhaps first conjectured on the basis of an insufficient number of particular cases, a general proposition comes to be more and more securely established by being connected with other truths through chains of inferences, whether consequences are derived from it that are confirmed in some other way or whether, conversely, it is seen to be a consequence of propositions already established. (BEG 5)

This suggests there is justification that can confirm a truth while stopping short of actually establishing it as a truth. This is also suggested in *Foundations*:

Induction itself depends on the general proposition that the inductive method can establish the truth of a law or at least some probability for it. If we deny this, induction becomes nothing more than a psychological phenomenon, a procedure which induces men to believe in the truth of a proposition, without affording the slightest justification for so believing. (FA 4)

There may seem to be a tension between Frege's beliefs that truth is the aim of science (see, for instance, PW 2) and that inductive reasoning does not license us to infer that our scientific theories are true. And one might be tempted to say that Frege is simply wrong about natural science. After all, Frege had very little to say about the natural sciences, and this tension could be regarded as symptomatic of the fact that Frege's real interest is in mathematics and not in the natural sciences. One might suspect that he just did not notice that his remarks could not actually be applied to the natural sciences. But it is important to notice that Frege's views on natural science are not entirely unreasonable. Let us look more closely at what Frege's views about logic commit him to saying about natural science.

First, Frege is committed to the belief that all scientific claims should ultimately be expressible in Begriffsschrift. This is important because logical laws must apply to all scientific claims. In particular, the use of deductive inference in scientific work is every bit as legitimate and necessary as inductive methods. Indeed, deductive inference is necessary if one is to employ inductive reasoning. He says (FA 23) that the laws established by induction are not enough. Furthermore, not only are deductive inferences legitimate but, also, no legitimate scientific method can violate Frege's laws of thought. A scientific method that, in practice, seemed to license both some statement and its denial would either be illegitimate or would not, in fact, license both. (It might, for instance, turn out that some of the presuppositions were false.) It is also important to note that, if this view were to conflict with actual scientific practice, Frege would not be likely to retreat. For he criticizes mathematical practice and deliberately advocates a mathematical method that conflicts with this practice.

But, of course, none of this marks Frege as naive about natural science or induction. In fact, inductive logic must be based on deductive logic. Carnap's attempt to construct an inductive logic in *Logical Foundations of Probability* is based on his introduction, in chapter 3, of a deductive logic and appropriate logical languages. And there are contemporary probabilists who are as committed as Frege to the importance of deductive inference in the natural sciences, to the view of truth as the aim of science, and to the view that inductive confirmation does not license detachment.[11]

One way to characterize the generality (FA 4) or transcendence of particulars (BEG 5) which Frege attributes to the laws of Begriffsschrift is to say that these laws are applicable to all scientific reasoning. The laws of thought must be universally applicable. How does Begriffsschrift exhibit this universal applicability?

Begriffsschrift contains two sorts of symbols: those that do and those that do not have a determinate meaning (*bestimmten Sinn*, BS 1). The symbols that have a determinate meaning are the logical symbols (e.g., the negation sign and the conditional sign.) The other symbols are variables. Thus, we can understand Frege's contrast, in *Foundations*, of the general logical laws with truths that contain "assertions about particular objects" (FA 4). The symbols that can stand for objects and that appear in the Begriffsschrift expression of laws are variable symbols that do not stand for any particular objects. Quantifiers binding these variables cannot range over some restricted universe, they must range over the unrestricted universe.[12] The universe contains all objects of all sciences. In particular, then, not only do the laws of Begriffsschrift not contain assertions about particular objects, they do not contain assertions that are restricted to particular scientific fields. Thus, the justification of a truth that can be inferred from the logical laws is not dependent on any particular facts. On the other hand, since the variables range over the entire universe, including the realm of the natural sciences, Frege's logical laws can be used in inferences concerning any objects, including objects that are the subject of some special science. In this sense, logic is the most general science.

This generality or universal applicability of logic is precisely what is

[11]See, for example, the corpus of Richard Jeffrey—in particular "Valuation and the Acceptance of Scientific Hypotheses," 237–246; "Probable Knowledge," 166–180; and "Dracula Meets Wolfman: Acceptance vs. Partial Belief," 157–185.

[12]The significance of Frege's insistence that quantifiers range over the unrestricted universe has been much discussed. My views have been influenced by Jean van Heijenoort's discussion in "Logic as Calculus and Logic as Language." I have benefited from correspondence and discussion with Cora Diamond and Thomas Ricketts. I have also benefited from reading the discussions in Diamond's "Frege against Fuzz" and Ricketts's "Objectivity and Objecthood: Frege's Metaphysics of Judgement" and "Generality, Meaning, and Sense." See also Leila Haaparanta, *Frege's Doctrine of Being*, chap. 2, and Sluga, *Gottlob Frege*.

needed in order for Frege to introduce a logical source of knowledge which can be tied to Kantian analyticity, as Frege understands it. For if the laws of thought—the *same* laws of thought—are applicable in every discipline, they cannot be a part of any special science. Furthermore, if these laws of thought—which are not part of some special science and are applicable everywhere—are the *only* tools, other than the definitions of concepts, involved in the proof of some truth, then it is not unreasonable to characterize the justification of this truth as consisting in just the twisting and turning (or analysis) of the concepts involved. This sort of characterization is also suggested by Frege's comments about the laws of identity in *Foundations*, where he says, "As analytic truths they should be capable of being derived from the concept itself alone" (FA 76).

Of course, one might still worry that the tools provided by the logical laws would constitute additional support. Even if the logical laws are everywhere applicable, the very notion of applicability seems to presuppose that appeals to these laws introduce something additional to the truths to which they are applied. But this is not quite correct. For example, Law (1) of Begriffsschrift is (in contemporary notation): $a \rightarrow (b \rightarrow a)$.[13] Frege gives the following informal gloss on the content of this law: "If a proposition a holds, then it also holds in case an arbitrary proposition b holds" (BEG 29). One of Frege's explanations, in "The Aim of 'Conceptual Notation'," of the difference between his logic and Boole's is that the formulae used to express Boole's laws "are actually only empty schemata" (CN 97). These empty schemata will not appear in proofs. Rather, in a proof one might use the claim that "A) If the sum of the angles of the triangle ABC is two right angles, this also holds in case the angle ABC is a right angle."[14] But A) does not require some sort of outside justification; its truth is evident from its content alone. The expression of A) in Frege's logical notation is simply supposed to make its content more perspicuous. The above expression of Law (1), of course, constitutes only a partial expression of A) in Frege's notation. There are, after all, concepts involved in A) (angle, triangle, etc.) to which general, but nonlogical, laws are applicable. However, no such law is necessary for the justification of A). This expression makes enough of the content of A) perspicuous that its truth is evident.

What is the source of this evidence of the logical laws, or the ultimate

---

[13]This is Law (1) in *Begriffsschrift* (BEG 29) and Law (I) in *Grundgesetze* (GGA 34).
[14]This is Frege's example; see BEG 29.

ground of their justification? It is tempting to suppose that some sort of metalogical argument, the definition of the conditional via exhibition of a truth-table, for instance, would provide the justification of the law discussed above. But Frege's view of logic precludes such justification. For Frege's logical notation is intended as a means for expressing, for *every* proposition, *all* content that is relevant to determining the correct inferences in which it can figure. Thus, if the justification of a logical law requires an argument in which the definition of the conditional figures, this definition must be expressible in Begriffsschrift. But no such definition is expressible in Begriffsschrift. Frege's symbol for the conditional, the conditional-stroke, is a primitive symbol in his language. Although Frege does introduce a definition for his conditional-stroke, it is not a definition in his notation. What, then, is the status of this definition? Frege says,

> We have already introduced a number of fundamental principles of thought in the first chapter in order to transform them into rules for the use of our signs. These rules and the laws whose transforms they are cannot be expressed in the ideography [Begriffsschrift] because they form its basis. (BEG 28/BS 25)

The upshot, it seems, is that there is no room for a substantive metaperspective.[15]

It may seem to follow that Frege's view requires not only the absence of a metaperspective for logic but also the impossibility of the justification of the laws of logic. But this is not quite right. Frege does not think that we are precluded from justifying all logical laws. Many logical laws can be proved from other logical laws. In Frege's introduction of each version of his Begriffsschrift (in *Begriffsschrift* and *Basic Laws*) he provides a short list of primitive logical laws from which the rest are to be proved. He says, "The question why and with what right we acknowledge a law of logic to be true, logic can answer only by reducing it to another law of logic. Where that is not possible, logic can give no answer" (BLA 15). Where that is not possible, of course, is any point at which a primitive law of logic is reached. It seems that, if logic must be universally applicable, the primitive laws of logic cannot be justified.

---

[15]The view that Frege had no room for a metaperspective has also been defended in G. P. Baker and P. M. S. Hacker, *Frege: Logical Excavations*; Leila Haaparanta, *Frege's Doctrine of Being*; Thomas Ricketts, "Objectivity and Objecthood: Frege's Metaphysics of Judgement."

For if logic can give no answer, surely none of the special sciences can give an answer.

But is this an unfortunate consequence? Should it be possible to justify a primitive law of logic? The easy answer is that, for Frege, justification must stop at some point and that the obvious place for justification to stop is when the primitive logical laws are reached.

I said, earlier, that I would argue that Frege is committed to a view about the ultimate ground of the justification of primitive logical laws and argue that it follows that Frege's antipsychologism holds for the issue of the ultimate grounds of primitive truths. If, as I have now argued, the primitive logical laws cannot be justified, they cannot be justified by any mental process. This is a part of my response to Kitcher's claim that the ultimate grounds of justification (or sources of knowledge) have to do, for Frege, with mental states or processes that produce beliefs.

There is something unsatisfying about this response. If no mental process can be the justification of a primitive logical law, might not something mental, nonetheless, constitute its ultimate ground? This may seem to be suggested by Frege's talk, in the above quotation from *Begriffsschrift*, of transforming fundamental principles of thought into rules for the use of his signs. My own use of the word 'thought' for the ultimate ground of the most general truths may seem to suggest this as well. It may seem, in particular, that the primitive logical laws are simply formalizations of laws in accord with which, given our nature, we must think. This looks to be a sense in which one might take mental processes of a certain sort of being (us) as the ultimate ground of the primitive logical laws.

Although Frege says little about the ultimate ground of the primitive logical laws in his early writings, there is a revealing discussion in the introduction to *Basic Laws*. Two passages from this discussion might be taken to support the above description of the ultimate ground of primitive logical laws. These passages deserve to be quoted in their entirety. He says:

> But what if beings were even found whose laws of thought flatly contradicted ours and therefore frequently led to contrary results even in practice? The psychological logician could only acknowledge the fact and say simply: those laws hold for them, these laws hold for us. I should say: we have here a hitherto unknown type of madness. (BLA 14)

Later in the same discussion he says:

If we step away from logic, we may say: we are compelled to make judgments by our own nature and by external circumstances; and if we do so, we cannot reject this law—of Identity, for example; we must acknowledge it unless we wish to reduce our thought to confusion and finally renounce all judgment whatever. I shall neither dispute nor support this view; I shall merely remark that what we have here is not a logical consequence. What is given is not a reason for something's being true, but for our taking it to be true. Not only that: this impossibility of our rejecting the law in question hinders us not at all in supposing beings who do reject it; where it hinders us is in supposing that these beings are right in so doing, it hinders us in having doubts whether we or they are right. At least this is true of myself. If other persons presume to acknowledge and doubt a law in the same breath, it seems to me an attempt to jump out of one's own skin against which I can do no more than urgently warn them. (BLA 15)

The possibility of supposing beings who judge in accord with laws of thought which contradict our laws of thought suggests that our belief in these laws is a feature of our psychological makeup and that Frege's project, in setting out the primitive logical laws, might be described as an attempt to describe the nature of our mental processes. For Frege seems to be acknowledging the possibility of beings who actually could understand the content of a primitive logical law, yet reject it. It would seem, then, that it is not quite right to say that the content of a primitive logical law suffices for its justification; rather, one should say that its content suffices for its justification *for beings like us*. Furthermore, the remark about jumping out of one's own skin suggests that our inability to justify these laws is a feature of our peculiar psychology. Perhaps we cannot justify these laws only because we are not capable of entertaining their denial. How seriously should this be taken?

I think that it would be a mistake to place too much weight on the supposition of the existence of such beings. For Frege says, in his discussion of the possibility of a being who rejects the law of identity, that our inability to reject something is not a reason for something's being true, but for our taking it to be true. Furthermore, if we look at Frege's views more closely, it becomes apparent that it follows from his views that we can do no more than abstractly suppose they exist. We cannot tell an intelligible story about their beliefs because there is a sense in which the content of an assertion cannot be divorced from the primitive logical laws.

Frege views Begriffsschrift, his logical language, not simply as a means for identifying correct inferences but primarily as a means for setting out the conceptual content of a judgement. For Frege, these tasks are not independent. In section 3 of *Begriffsschrift* Frege explains his reasons for not distinguishing, in his logical language, between the subject and predicate of a judgement. He indicates that there may be a difference between judgements that are expressed by sentences having different subjects but for which "the consequences derivable from the first, when it is combined with certain other judgments, always follow also from the second, when it is combined with these same judgments" (BEG 12). But, whatever differences there may be in the content of these judgements, there is also something common to their content. Frege calls what is the same in such judgements the "conceptual content" (*begrifflichen Inhalt*, BS 3). Frege's language is designed to represent the features of a judgement in such a way as to make clear in exactly which correct inferences the judgement can figure. To describe this as the conceptual content of the judgement is to suggest that our understanding of the content of the judgement already includes some sort of correct understanding of the logical laws. For instance, someone who understands the content of the claim that nobody read Frege's *Begriffsschrift* must know that it is incorrect to infer that there was someone (namely, nobody) who read *Begriffsschrift*. In fact, Frege's explanations of how to translate everyday sentences into Begriffsschrift expressions rely on our sharing an understanding of what inferences are correct. As a result, on Frege's formulation, as on Kant's, the claim that a truth is analytic is a claim about its content.

Frege says we can suppose beings who reject the law of identity, that is, who reject the claim that every object is identical to itself. What would it be like to be confronted with such a being? Presumably this being would say that not every object is identical to itself. But the location of some being who says this would not be likely to result in a claim that we have found a being who rejects the law of identity. We would certainly, as Frege says, be hindered in supposing the being to be right. But how we would describe the nature of her, his, or its error? If we were simply to assume that the being is denying the law of identity, we would be hindered in far more than our ability to suppose the being is right. As Frege says, much later, "The assertion of a thought which contradicts a logical law can indeed appear, if not nonsensical, then at least absurd; for the truth of a logical law is immediately evident of itself, from the sense of its expression" (CP 405). Given the apparent absurdity or nonsensical nature of the being's explicit claim, it is not clear how we could communicate with this being. Exchanges of absurdities would not constitute communication. In order to communicate, it would be necessary to interpret this being's assertion as an intelligible, nonabsurd thought. The content of such a thought cannot, on Frege's view, be independent of our logical laws. We might, for instance, understand this being to be using the expression "is identical to" for some relation other than identity. But on this reinterpretation of the being's assertion, we would be attributing a different content to the being's apparent denial of the law of identity. An understanding of this different content, moreover, would include an understanding of how the thought could be correctly used in inferences. That is, the attribution of this thought to the being in question would involve the attribution of our logical laws. The primitive logical laws will constrain how we can understand this being's assertions and, as a result, we will be unable to give an account of the content of the assertions of a being whose laws of thought contradict ours.

It may seem to follow that we *cannot* suppose the existence of beings whose laws of thought contradict ours. After all, I have argued that Frege's view would commit us to understanding any assertion as an assertion that does not contradict primitive logical laws. How, then, are we to make sense of Frege's saying, "But what if beings were even found whose laws of thought flatly contradict ours and therefore frequently led to contrary results even in practice?" (BLA 14). I have not argued that Frege's view guarantees the success of our attempts to make sense of such a being's assertions. I have argued only that, given Frege's view, our ability to communicate with someone is dependent on our ability to succeed in understanding her or his assertions as assertions that do not obviously contradict primitive logical laws. Suppose, however, we were confronted with a being whose utterances defied such interpretation. In such a case, we would be able to make no sense of this being's utterances. I am not sure that we would be inclined to say, in such a case, that the being is actually judging in accord with laws of thought which contradict ours. More likely we would say, as Frege says *he* would say: "We have here a hitherto unknown type of madness" (BLA 14).

It is important to notice, however, that it does not follow that it is impossible to make an incorrect inference from some truth, once one has understood its content. Nor does it follow that anyone who makes an incorrect inference is mad. Conceptual content can be sufficiently complicated that one can make mistakes. Frege indicates in a late unpublished paper, "Logic in Mathematics," that it might be psychologically impossible for one to keep the complete definition of the integral in mind during all reasoning about integrals (PW 209). One might, as a result of ignoring some detail of the definition of the integral, make a mistaken inference. Furthermore, it might turn out that, for some people, some such mistake would be uncorrectable because their limited abilities would preclude them from simultaneously entertaining both the relevant detail and the inference in question. These people would not be mad. On the other hand, these people would not be right—their error should, if Frege's Begriffsschrift allows the expression of the definition of the integral, be demonstrable. The point is that once one has identified and, using Frege's Begriffsschrift, represented the conceptual content of all the propositions involved in an inference, one has all the tools necessary to determine whether or not the inference is legitimate.

Thus, although Frege does not explicitly attempt to answer "why and with what right" we take the laws of logic to be true, he does the best that can be done. Insofar as the question of with what right we take the laws of logic to be true is a question about the justification of these laws, the answer to the question can be found in the content of the primitive laws on which they depend. Nonprimitive laws can be proved from primitive laws. Primitive logical laws need not and cannot be justified. Any account of the impossibility of our rejecting a primitive logical law, Frege says, is "not a reason for something's being true, but for our taking it to be true" (BLA 15). All Frege can do is try to make the conceptual content of the primitive laws clear. Frege's writings do, however, provide an explanation of why, given the nature of the laws of logic, the question of the justification of the primitive laws of logic is not a proper question. For the conceptual content of a judgement is a part of its content. The upshot is that the primitive laws of logic must be inextricably bound up in our understanding of the content of our judgements. This is, perhaps, why Frege says, in a footnote in *Foundations*, "Observation itself already includes within it a logical activity" (FA 99). On the other hand, the question of why we take the laws of logic to be true, or what mental processes a person uses to convince herself of this, is a psychological question. It is not improper to attempt to answer such a question, but it is a mistake to assume that the answer to this question will tell us anything about the justification of the laws of logic. Indeed, the logical laws apply to all scientific research, including psychology. An understanding of the content of the questions asked about our psychology cannot be divorced, any more than the content of any other thoughts, from the logical laws.

To say that thought is the ultimate ground of the justification of logical truth, for Frege, is not to say anything about the mental or psychological process by which one can come to know a logical truth. Although the justification of a primitive logical law is evident from its content, it does not follow from this that the act of considering the content of a primitive logical law constitutes its justification. Justification, in the sense with which Frege is concerned, is not a feature of the person who believes the truth. It does not seem unreasonable to suppose that, in general, to make a claim about what is the ultimate ground of the justification of some truth is to say something about its justification in a nonmental sense. Frege does seem to think (see FA 17–21) that we convince ourselves of the truth of primitive truths of pure intuition (i.e., the primitive synthetic a priori truths) via the construction of mental pictures. But it is no more obvious that this mental construction constitutes the justification of these truths than it is that the act of considering the content of a primitive logical law constitutes *its* justification.

What does constitute the justification (or ultimate ground) of a primitive truth that Frege would categorize as synthetic a priori? With what right do we believe these to be true? Frege's only answer is to say the ultimate ground of such a truth must include inner intuition. He also says, "We are all too ready to invoke inner intuition, whenever we cannot produce any other ground of knowledge. But we have no business, in doing so, to lose sight altogether of the sense of the word 'intuition'" (FA 19). Frege does give us a means for recognizing actual invocations of inner intuition. These are found whenever the axioms of Euclidean geometry are needed for some proof. But although he always categorizes the axioms of Euclidean geometry as primitive truths whose ground is inner intuition, Frege says little about what inner intuition is. I suspect the explanation is that Frege took Kant to have answered this question as well as the corresponding question for synthetic a posteriori truths.

It should not be surprising if this is an unsatisfying answer—if it suggests that there may, after all, be no content to be found in Frege's writings for the threefold division he seems to draw. For one of the reasons for understanding Frege's project as an attempt to contribute to a version of Kantian epistemology is that Frege's explicit motivation for his project of proving the truths of arithmetic from primitive truths is incomplete. The advantage of understanding Frege's project as a contribution to a version of Kantian epistemology is that it explains both

Frege's claim that his proofs determine the epistemological nature of the laws of arithmetic (see, e.g., BLA 3) and the absence of an exposition of his own epistemological views.

It may no longer seem obvious that Frege's motivation for defining the numbers, on this reading, is epistemological. The issue of whether or not it is epistemological will depend on one's view of epistemology. Kitcher thinks that epistemology concerns the mental processes that produce knowledge—an issue in which, I have argued, Frege has no interest. There is no indication that Frege had any interest in the psychological processes by which we convince ourselves of truths or in the circumstances under which a person's true belief counts as knowledge. Thus, for Kitcher, Frege's interests cannot be viewed as epistemological. On the other hand, Frege's understanding of epistemology is not Kitcher's. On Frege's view, justification is not a mental process. The justification of a truth is independent of anyone's reason for believing it. To examine the justification of some truth whose grounds are inferential is to examine the argument for it. The justification of such a truth of some natural science is part of that science. But not all justification is inferential. Epistemology concerns the grounds of truths whose justification is not inferential. The fact that such a truth cannot be justified by an argument belonging to some scientific field is only an indication of its primitive nature, not an indication that the ultimate ground for its justification is some special sort of mental process.

Frege's arguments against psychologism are meant to be arguments against Kitcher's view of epistemology as well. On Frege's view, there are certainly psychological processes that go on when we reason correctly, but investigation into these processes is psychology, not epistemology.

This completes the argument for my interpretation of the motivation of Frege's project. On this view, Frege saw a problem with Kantian epistemology (as he understood it). This was that Kant was wrong about what the laws of thought were. Frege develops, in his *Begriffsschrift*, a means for setting out the true laws of all thought. And the result is that, on Frege's view, the serious Kantian will be obliged to substitute Frege's logic for traditional logic in the epistemological picture. But, once this is done, it becomes clear that the analytic/synthetic distinction must be reformulated. Frege does reformulate the analytic/synthetic distinction, but his reformulation, in turn, muddies the issue of whether the truths of arithmetic are analytic or synthetic.

From Frege's point of view, however, this muddying is a desirable result. Frege also seems to have had antecedent reasons for believing that Kant's claim that the truths of arithmetic are synthetic a priori may be wrong. For Frege's threefold division of truths according to their ultimate grounds is also viewed as a division of truths according to the domains over which they are applicable. He says, "Empirical propositions hold good of what is physically or psychologically actual, the truths of geometry govern all that is spatially intuitable, whether actual or product of our fancy" (FA 20). These propositions are all synthetic. Empirical truths are synthetic a posteriori and the truths of geometry are synthetic a priori. Analytic truths, Frege suggests, govern the widest domain of all—everything that fails to violate the logical laws, that is, everything thinkable.

If Frege's division is viewed in this way, to claim that arithmetic is synthetic a priori is to relegate it to the status of a special science, that is, to limit its range of applicability. Yet arithmetic is generally believed to apply, as Frege suggests logic must, to all thought. And, if this is true, the laws of number should have the same status as the laws of all thought—they should be analytic. Of course, none of this is convincing on its own. But some of Frege's other assumptions make it seem more reasonable to suppose that the laws of number might be analytic. Frege was clearly antecedently convinced that the inferences licensed by Aristotelian logic did not exhaust the inferences licensed by the laws of thought. That is, Frege believed that the laws of all thought licensed more than Kant believed they licensed. Frege also was antecedently convinced that the apparent invocations of intuition in descriptions of mathematical justification by Kant, by other philosophers, and by mathematicians did not really play any substantive role in the justification. It might not seem unreasonable to suspect, therefore, that these apparent invocations of intuition were actually glosses used to cover up the fact that certain inferences were not well enough understood. The successful expression of the true laws of thought, then, might provide details that would obviate the need for these glosses and show that arithmetic can in fact be derived from logic alone.

I have argued that, in order to understand Frege's demand for proofs of the simplest truths of arithmetic, it is necessary to suppose that Frege viewed himself as working within a philosophical tradition that provided an account of the available ultimate grounds of justification. Frege argues that it is not clear what the ultimate ground of the justification of the truths of arithmetic is. Since the ultimate ground of a truth is determined by the ultimate ground of the primitive truths from which it can be proved, and since the ultimate ground of a primitive truth is evident, Frege wants to find the primitive truths from which the truths of arithmetic can be proved. But if my interpretation, at this point, explains why Frege wanted these proofs, it does not explain what criteria determine whether a proof is acceptable for Frege's purposes. Although it is clear that, in order to prove the simplest truths of arithmetic from primitive truths, the numbers must be defined, two central questions remain unanswered. First, what is a primitive truth? Without an answer to this question, it will be difficult to be certain whether a proof of a truth of arithmetic is a proof from primitive truths or not. I have given one example of a primitive truth, Law 1) from Frege's *Begriffsschrift*, and explanations of how we can be certain of the ultimate ground of primitive logical truths. But it is not clear that this is sufficient. Second, what would make a definition of the numbers acceptable? I will begin my answers to these questions with an examination of Frege's notions of definition and definability.

CHAPTER 3

A Systematic Science

I T is now possible to put together a story that motivates Frege's project of proving the basic truths of arithmetic. On Frege's view, these proofs will enable him to locate, within a modified Kantian framework, the source of justification of the truths of arithmetic. The need for both locating the source of this justification and modifying the Kantian framework arises from Frege's antecedent conviction that intuition, in Kant's sense, is not required for the justification of the truths of arithmetic. Frege's explicit modification is the introduction of a substantive logical source of knowledge. The claim that there is a substantive logical source is made plausible by the replacement of Aristotelian logic with his new logic. On Frege's account, a truth either is a logical truth—that is, its justification depends solely on definitions and the laws of logic— or requires support from intuition. In order to support his antecedent conviction that intuition is not necessary for the justification of the truths of arithmetic, Frege must exhibit the justification of these truths.

The epistemological status of arithmetical truth, that is, the source of the justification of arithmetical truth, will depend on the status of a group of basic arithmetical truths from which all other truths of arithmetic can be proved. What are these basic arithmetical truths, and what is their epistemological status? Given Frege's exposition of his logical system, it is not obvious that the claim that $0 \neq 1$, which looks to be a basic arithmetical truth, is a logical truth. But, as Frege argues in chapter 1 of *Foundations*, it is by no means obvious that the claim that $0 \neq 1$

requires intuition for its support. Thus, since it is not immediately clear what the ground of the justification of the claim that $0 \neq 1$ is, it seems that it is not a primitive truth that $0 \neq 1$. The justification of the claim must have some structure to it. There must be some more basic truths that underlie its proper justification, some truths from which it can be inferred, and its status will be determined by the status of these more basic truths. Since mathematical justification is proof, Frege infers that it must be provable that $0 \neq 1$. The proof of this truth as well as of the other apparently basic, but not epistemologically primitive, truths of arithmetic will be required in order to determine the ultimate grounds of these truths.

Frege's epistemological motivation places strict constraints on what sorts of proofs will be of use. Set theoretic proofs of the basic truths of arithmetic, for example, will not help. Such proofs would simply allow one question to be traded for another. Instead of asking whether the basic truths of number are immediately justified via intuition or are immediately recognizable as logical truths, it would be necessary to ask this question about the basic truths of set theory. The answer to this question, however, is no more obvious than the answer to the original question. For the basic set theoretic truths are not obviously logical truths (there is no notation for sets in Frege's Begriffsschrift[1]), nor are they immediately justified via intuition. Frege says, "The terms 'multitude', 'set' and 'plurality' are unsuitable, owing to their vagueness, for use in defining number" (FA 58). Thus, not all mathematical proofs of the basic truths of arithmetic will satisfy Frege. In his *Begriffsschrift*, Frege has provided information that determines whether or not passage from one step to another in a proof is legitimate. He has also provided some information that will help determine whether or not a proof is finished, that is, whether or not its premises are epistemologically

[1]In *Begriffsschrift* itself, there is no notation either for sets or for extensions of concepts. Frege later amends this notation to include names for extensions of concepts. However, throughout most of his career, Frege takes the notion of an extension of a concept to differ from that of a set. The term 'set' as used by mathematicians is, on Frege's view, not sufficiently precise. See, for instance, BLA 29–32, PW 181–183, and FA 38–39. There is at least one unpublished comment, however, in which Frege actually characterizes his original project as an attempt to construe numbers as sets (*Mengen*; PW 269/NS 289). This appears in a short article written in 1924 or 1925. I think the reason for this deviation from his characterizations elsewhere is that one upshot of the nature of his failure in *Basic Laws* is that, contrary to his earlier claims, the notion of the extension of a concept was not clear at all. Thus, from this later perspective, it is not at all clear that there really was a distinction between Frege's notion of extension of a concept and the mathematical notion of set.

primitive. The underived logical laws introduced in *Begriffsschrift* are primitive.

It may seem, however, that, without a more elaborate account of what it is for a truth to be primitive, Frege cannot give a general account of what it is for a truth to stand in need of proof and, hence, that there are no general criteria for determining whether a proof of an arithmetic truth is finished (i.e., that it is a proof from primitive truths). And it might be tempting to infer that Frege has provided no criteria that will justify his claim to have proved the truths of arithmetic. However, this does not follow. For, if he is right about the analyticity of arithmetic, Frege need not give any more of an account of primitiveness or un-provability. Although no general account of primitiveness has been given, some examples of primitive truths, the primitive logical laws, are available. Thus, Frege might be able to give recognizable proofs of all the truths of arithmetic from recognizably primitive truths. In this case, no general account would be required. At first glance this seems to be exactly what Frege is doing. His aim is to show that the truths of arithmetic are analytic, and he tries (beginning with *Foundations* and finishing in *Basic Laws*) to show that all truths of arithmetic can be proved using only the general laws of logic and purely logical defini-tions of the numbers. Since *Begriffsschrift* clearly demarcates the bound-aries between logic and everything else, such a proof is easily identifi-able.

The role of proof in Frege's project seems quite straightforward now. However, there remain puzzles concerning the nature of Frege's defini-tions. The demand for definitions of the numbers is introduced with the opening question of *Foundations* (What is the number one?). Frege's initial discussion of his question seems to indicate that he is interested in saying precisely what it is we are talking about when we use the symbol '1'. And an investigation of Frege's motivation reveals that serious epistemological constraints dictate what can and cannot count as an answer. In particular, the definitions must be usable in proofs, for Frege's aim is to prove the truths of arithmetic from epistemologically primitive truths. These descriptions of Frege's definitions may seem straightforward, but some of their consequences are puzzling. If the definition of the number one is to be regarded as making precise what meaning had previously been associated with the symbol '1', it seems obvious that the symbol '1' is to be regarded as having had meaning prior to Frege's work. It seems to follow that the definition is not an arbitrary stipulation that assigns a meaning to a previously empty term.

This view is further supported by Frege's comment after his attempt to define "the Number which belongs to the concept F." He says, "That this definition is correct will perhaps be hardly evident at first. For do we not think of the extensions of concepts as something quite different from numbers?" (FA 80). Thus, it seems that Frege regards a definition as something that makes a substantive assertion and requires justification. On the other hand, Frege also says, "The definition of an object does not, as such, really assert anything about the object, but only lays down the meaning [*setzt die Bedeutung*] of a symbol" (FA 78). This seems to contradict the other passage. It also suggests that Frege is regarding the symbol '1' as having had no meaning prior to his work. For if Frege's definition is simply a precise description of the meaning '1' already has, then the definition does assert something about the object—it tells us, for instance, that the number one is really an extension of a concept. Given the central role played by definition in Frege's project, this looks like a serious tension.

It may be tempting to regard the second passage as a mistake on Frege's part. However, if it is a mistake, it is not an isolated mistake. For, earlier in *Foundations*, before Frege begins to describe the actual definitions he will use, he says, "It must be noted that for us the concept of Number has not yet been fixed" (FA 74). And some of the definitions Frege discusses are clearly intended to be regarded as arbitrary stipulations of the meaning of a term. For instance, when he introduces a definition of the manufactured term *gleichzahlig*, he says, "I must ask that this word be treated as an arbitrarily selected symbol, whose meaning is to be gathered, not from its etymology, but from what is here laid down" (FA 79). If Frege's project is to be given a coherent explanation, it will be necessary to determine whether or not he regards definitions as arbitrary stipulations of meaning to previously meaningless terms and to provide a convincing reading of the passages that appear to suggest otherwise. In order to do this, it will be helpful to examine some of the other discussions of definition which appear in *Foundations*.

Although Frege gives no explicit account of definition in *Foundations*, he does make a number of remarks about definitions and defining. In particular, three criteria of adequacy for definitions emerge from the discussions in *Foundations*. These are:

1. definitions must not lead to contradictions;[2]

[2]See FA ix.

2. definitions must be usable in mathematical proofs (they must be "fruitful");[3] and
3. a definition must precisely delimit the term defined (a term, once defined, must have "sharp boundaries").[4]

A careful examination of each of these criteria will make the nature of Frege's understanding of "definition" considerably clearer.

The first seems the most obvious and least exceptionable. After all, Frege wants to use his definitions in proofs, and he can hardly expect to use a principle that will lead to contradictions in his proofs. Frege also criticizes mathematicians for failing to justify their definitions (FA ix). It seems natural to interpret this remark as an indication that Frege believes that it is necessary to *prove*, for each definition, that it will not lead to a contradiction. But what would such a proof be like?

We are accustomed, today, to viewing logical axioms and rules of inference as constituting parts of formal theories and to viewing formal theories as mathematical structures whose features can be the subject of mathematical proof. To prove that a definition can be introduced in a formal theory without allowing the proof of a contradiction in the theory is not to give a proof within that theory but, rather, to give a proof about the theory, a metatheoretic proof. It is perfectly natural, from our perspective today, to view Frege's logic as a formal theory and to view the requirement that a definition not lead to contradiction as something that can be shown only via a metatheoretic consistency proof. Did Frege require metatheoretic consistency proofs? I have argued that Frege's epistemological presuppositions preclude him from regarding his *Begriffsschrift* as setting out a formal theory. Had Frege required that definitions be justified via such consistency proofs, this would create a serious tension either in my reading or in Frege's project. There is, however, no evidence that Frege's consistency requirement on definition is a demand for a metatheoretic consistency proof.

Although Frege introduces definitions in both his pre-*Foundations* *Begriffsschrift* and his post-*Foundations* *Basic Laws*, none of these definitions is preceded or followed by an attempt at a consistency proof. It does not follow, of course, that consistency proofs are not required. The definitions in *Begriffsschrift* clearly are meant to be nothing more nor less

[3]See FA viii, 34, 81.
[4]The content of the sharp boundary requirement is most explicitly set out for concept expressions on FA 87 and for object expressions on FA 73. See also FA 8, 68, 78, 112, 113, 115.

than arbitrary stipulations.[5] It seems possible that the definitions Frege is talking about in *Foundations* are meant to have a different status and that the criteria of adequacy for definitions introduced in *Foundations* need not be applied to these definitions. But Frege's remarks in *Foundations* about definitions clearly *are* meant to apply to the definitions he will give in *Basic Laws*. What of the absence of consistency proofs for definitions in that work? One might suspect that in 1884, when *Foundations* was published, Frege believed proofs were necessary to establish the consistency of definitions, but that he later came to believe that this was wrong. Such an account, however, attributes to Frege a conscious change of mind about the nature of definitions. In his later work, Frege explicitly denies having changed his views on the nature of definitions. Frege's most detailed discussions of definition in his later work appear in his criticisms of Hilbert's work on the foundations of geometry. One of Frege's central objections to this work is that Hilbert's understanding of 'definition' is not the standard mathematical one. In each of Frege's two papers on the foundations of geometry, he contrasts Hilbert's use of the term "definition" with his own use, which he claims fits with the traditional use (CP 275). In fact, the second of these papers contains the following rather hysterical remark:

> Whoever wilfully deviates from the traditional sense of a word and does not indicate in what sense he wants to use it, whoever suddenly begins to call red what otherwise is called green, should not be astonished if he causes confusion. And if this occurs deliberately in science, it is a sin against science. (CP 300)

It seems clear, from this remark, that if Frege's views on definition changed between 1884 and 1906 (the year in which the above passage was published), he was unaware of the change.[6]

---

[5]See BEG 55–56.

[6]Another significant reason for thinking that Frege's views on definition did not change is that, taken together, his early and late remarks on definition form a coherent picture. There is no apparent tension between the remarks on definition in *Foundations* and later works. In fact, remarks about definition in *Foundations* are repeated and amplified throughout Frege's later writings. Compare, for instance, "The definition of an object does not, as such, really assert anything about the object, but only lays down the meaning [*Bedeutung*] of a symbol. After this has been done, the definition transforms itself into a judgement" (FA 78) with the following remark in a letter to Hilbert in 1899: "Every definition contains a sign [expression, word] which had no meaning before and which is first given a meaning by the definition. Once this has been done, the definition can be turned into a self-evident proposition which can be used like an axiom" (COR 36) and

The absence of consistency proofs for his definitions is not the only evidence that Frege did not require these proofs. In fact, Frege nowhere indicates that the consistency requirement for definitions forces one to prove that definitions are acceptable. Furthermore, in a discussion of traditional mathematical strategies for defining certain notions in the second volume of *Basic Laws*, Frege mentions the possibility that this strategy might lead to contradictory definitions. He mentions that proofs of consistency are not given in these cases. Then he says,

> In general, we must reject a way of defining that makes the correctness of a definition depend on our having first to carry out a proof; for this makes it extraordinarily difficult to check the rigour of the deduction, since it is necessary to inquire, as regards each definition, whether any propositions have to be proved before laying it down. (TWF 143)

Thus, it is difficult to imagine that Frege viewed his consistency requirement as a requirement that the introduction of a definition be preceded by a consistency proof. How, then, is his consistency requirement to be understood?

When he discusses the consistency requirement directly, in his first essay on the foundations of geometry, he says, "We must set up such guidelines for giving definitions, that no contradiction can occur. Here

---

with the following remark from the first of his essays on Hilbert: "In mathematics what is called a definition is usually the stipulation of the meaning of a word or a sign. A definition differs from all other mathematical propositions in that it contains a word or sign which hitherto has had no meaning but which now acquires one through it" (CP 274). Frege also says virtually the same thing in CP 139, 294, 315, and in PW 208, 240, 244, 248.

Similarly, Frege's objection to a preliminary formulation of a definition of the concept number (FA 68) because it will not allow us to decide whether or not Julius Caesar is a number is echoed throughout his essays on the foundations of geometry. An example is Frege's assertion in the second essay (CP 303) that a definition of the concept of point must determine, for each object (including his pocket watch), whether or not it is a point. Other passages in which Frege appears to repeat this can be found in TWF 139 and PW 229, 241. The apparent tension in Frege's notion of definition is not between *Foundations* remarks on definition and later remarks but, rather, between the notion of definition which appears throughout his writings and the role of definition in the project described in *Foundations*.

It should also be noted that the existence of an unpublished paper from the late 1890s titled "The Argument for my stricter Canons of Definition" does not provide evidence that Frege's views on definitions changed. The canons discussed here are not stricter than Frege's earlier canons but, rather, are stricter than those employed by mathematicians. Also, the argument for these canons presumes that mathematicians agree with Frege about the nature of definitions.

it will essentially be a matter of preventing multiple explanations of one and the same sign" (CP 275). The actual constraints that are imposed by the consistency requirement are discussed in detail when he prepares the reader, in *Basic Laws*, for his definitions of the numbers. In section 33 of the first volume of *Basic Laws*, Frege introduces seven principles of definition, three of which are specifically introduced as constraints designed to ensure the consistency of his definitions. One of these (Principle 2) is that a term not be defined twice; another (Principle 3) is as follows: "The name defined must be simple; that is, it may not be composed of any familiar names or names that are yet to be defined" (BLA 90). Principle 4 also states that the only sort of term which can be defined is a "simple sign, not previously employed" (BLA 91). It should be clear, then, that Frege's consistency requirement on definitions is not a requirement that the introduction of a definition be preceded by a consistency proof.

This interpretation of Frege's consistency requirement supports the view of definitions as arbitrary stipulations. For the principles of definition in *Basic Laws*—particularly the principle that requires that a term that is to be defined not have been previously employed—appear to guarantee that definitions can amount to nothing more than arbitrary abbreviations. This view is also supported by the following claim in Frege's first essay on the foundations of geometry: "No definition extends our knowledge. It is only a means for collecting a manifold content into a brief word or sign, thereby making it easier for us to handle. This and this alone is the use of definitions in mathematics" (CP 274). And it is supported by this remark, also from the first essay on the foundations of geometry: "Never may something be represented as a definition if it requires proof or intuition to establish its truth" (CP 275). All these remarks fit with Frege's denial, in *Foundations*, that a definition asserts something about an object. The puzzle of how to understand his suggestion that definitions require justification remains, however, and is joined by a new puzzle. He says in *Basic Laws* that a previously employed symbol cannot be defined, and yet a central aim in that work is to define the symbol '1'—which clearly *was* employed prior to Frege's work. I will turn to this presently.

The second requirement on definitions which emerges from the *Foundations*, the fruitfulness requirement, seems, at first glance, as obvious and unexceptionable as the consistency requirement seemed initially. On closer examination, this requirement also seems to support the view that definitions can be nothing more than arbitrary abbreviations. The requirement first appears in the introduction to *Foundations*, where he says, "Even I agree that definitions must show their worth by their fruitfulness: it must be possible to use them for constructing proofs" (FA ix). Later on he says, "Definitions show their worth by proving fruitful. Those that could just as well be omitted and leave no link missing in the chain of our proofs should be rejected as completely worthless" (FA 81). It is not difficult to see that the fruitfulness requirement must hold on Frege's definitions. After all, Frege's reason for wanting definitions of the numbers is to prove the truths of arithmetic from logical laws and definitions. Thus, if he is to succeed, his definitions must be usable in proofs, that is, they must satisfy the fruitfulness requirement. Of course, it does not follow that it is reasonable to extend this requirement to all definitions. In particular, it seems odd to require this if one understands definitions to be usable in contexts other than mathematics. However, it is not entirely clear that there was any point in his career at which Frege considered nonmathematical definitions. It is certainly true that his definitions of the numbers are meant to be mathematical and to be used in proofs. It seems that it should do no harm to assume that Frege is talking about mathematical definitions in the passages quoted above. It should be noted, also, that if we understand the fruitfulness condition discussed in these passages to be a condition on mathematical definitions, then Frege adhered to this requirement as late as 1919. For, in some notes written in 1919, Frege says, "A definition in arithmetic that is never adduced in the course of a proof, fails of its purpose" (PW 256).

Frege's concern with rejecting definitions that are worthless might be taken to indicate that he wants to distinguish substantive from nonsubstantive definitions. The suggestion, particularly of the later passage quoted, seems to be that a definition has worth only if its use is *necessary* for some proof.[7] This reading of the passage in question, however, is mistaken. In order to realize that this reading must be mistaken it suffices to recall Frege's example, in *Foundations*, of a fruitful definition. This is the definition of the continuity of a function (FA 100).

The definition of the continuity of a function is not necessary for any mathematical proof. For this definition amounts simply to the statement that 'f is continuous at x_0' is to be true just in case

[7]In "Frege: The Last Logicist" (LL), Paul Benacerraf has argued that this is Frege's position.

$$\lim_{x \to x_0} f(x) = f(x_0).$$

Now it should be clear that any theorem in whose proof this definition figures can also be proved without use of the definition. Consider, for instance, the intermediate-value theorem. This theorem can be stated as follows: Suppose that f is a real-valued function that is continuous on the interval $[a, b]$. Suppose also that $f(a) \neq f(b)$ and that Y is any number between $f(a)$ and $f(b)$. Then there exists a number X such that $a < X < b$ and $Y = f(X)$. Now since the assumption that f is continuous on $[a, b]$ is necessary, and since the notion of continuity is not primitive, the definition of continuity figures in the proof of this theorem. And the above statement of this theorem could not be proved without using this definition. However, it does *not* follow that the intermediate-value theorem could not be proved without the definition of continuity, for the theorem could be restated without any use of the word 'continuous'. It would be necessary only to substitute, for the condition that f is continuous on $[a, b]$, the condition that for any number c, in the interval $[a, b]$,

$$\lim_{x \to c} f(x) = f(c).$$

It should be clear that a proof of this version of the intermediate-value theorem would require no use of the definition of continuity. Similarly, any other theorem in which the word 'continuous' appears can be restated so as to be provable without any use of the definition of continuity. Thus, Frege cannot be requiring that a definition's worth is dependent on its being *necessary* for some proof. Furthermore, Frege says:

> In fact it is not possible to prove something new from a definition alone that would be unprovable without it. When something that looks like a definition really makes it possible to prove something which could not be proved before, then it is no mere definition but must conceal something which would have either to be proved as a theorem or accepted as an axiom. (PW 208)

As in the discussion of Frege's consistency condition, the characterization of definition which seems to emerge from Frege's fruitfulness condition is that of an arbitrary abbreviation. It seems, then, that the worth of mathematical definitions is purely pragmatic. Frege's fruitfulness requirement simply amounts to saying that whatever mathematicians find useful for constructing proofs is fruitful.

At this point, there seems to be another mystery surrounding the fruitfulness condition. This mystery is that Frege bothers to state it at all. What does Frege mean to be ruling out as a possible definition? Frege gives one example in the introduction to *Foundations*. He says, of the writing in mathematical texts,

> When the author feels himself obliged to give a definition, yet cannot, then he tends to give at least a description of the way in which we arrive at the object or concept concerned. These cases can easily be recognized by the fact that such explanations are never referred to again in the course of the subsequent exposition. For teaching purposes, introductory devices are certainly quite legitimate; only they should always be clearly distinguished from definitions. (FA viii)

Some years later, in an essay titled "Logic in Mathematics," Frege makes a similar statement.

> When we look around us at the writings of mathematicians, we come across many things which look like definitions, and are even called such, without really being definitions. Such definitions are to be compared with those stucco-embellishments on buildings which look as though they supported something whereas in reality they could be removed without the slightest detriment to the building. We can recognize such definitions by the fact that no use is made of them, that no proof ever draws upon them. (PW 212)

Many mathematical textbooks begin with discussions of what numbers are (or, in geometry textbooks, what points are). One purpose of Frege's mentioning the fruitfulness condition is to distinguish between his explicit definitions and these introductory devices.[8]

The third criterion of adequacy for definitions—the requirement that a definition fix sharp boundaries—is the most substantive. To understand what this requirement comes to, it is necessary to consider definitions of concept-words separately from definitions of object-words. The demand that a definition of a concept-word fix sharp boundaries amounts to the demand that the definition determine, for each object, whether or not it falls under the concept defined. Thus, Frege exhibits

---

[8]Another purpose may be to distinguish definitions from hints or elucidations. See, for instance, CP 299–302.

the inadequacy of a preliminary definition of the concept of number by noting that it will not make it possible to decide whether or not Julius Caesar is a number (FA 68). Frege says:

> All that can be demanded of a concept from the point of view of logic and with an eye to rigour of proof is only that the limits to its application should be sharp, that it should be determined, with regard to every object whether it falls under that concept or not. (FA 87)

For object-words, a definition must pick out exactly one object. Frege says, "If we are to use the symbol a to signify an object, we must have a criterion for deciding in all cases whether b is the same as a, even if it is not always in our power to apply this criterion" (FA 73). Thus, a concept can be used for defining an object just in case it is possible to show:

1. that some object falls under this concept;
2. that only one object falls under it. (FA 88)

As with the other criteria of adequacy, Frege continues to adhere to this criterion throughout his career. In his discussion in the preface to *Basic Laws* he says, about the strategy of using a property to define zero, "Only when we have proved that there exists at least and at most one object with the required property are we in a position to invest this object with the proper name 'zero'" (BLA 12).[9] And he says, in a discussion of definition from volume 2 of *Basic Laws*, "A definition of a concept (of a possible predicate) must be complete; it must unambiguously determine, as regards any object, whether or not it falls under the concept (whether or not the predicate is truly ascribable to it)" (TWF 139).[10] It is worth noting that this criterion of adequacy for defini-

[9]It is important to note that, although the definition of zero will ultimately be used in formulating a definition of the number one, the use of the word 'one' in the statement of this criterion of adequacy for definitions does not in any way make Frege's definition of the number one (or its justification) circular. To say that, for instance, a description, D, picks out exactly one object is simply to make a claim of the following logical form: $(\exists x)[Dx \ \& \ (y)(Dy \leftrightarrow y = x)]$. Thus, the condition of adequacy can be stated without any use of the word 'one'.

[10]It should be emphasized that it does not follow that all knowledge is based on definitions. Frege writes: "The question of whether a given stone is a diamond cannot be answered by the mere explanation of the word 'diamond' itself. But we can demand of the explanation that it settle the question objectively, so that by means of it everyone well acquainted with the stone in question will be able to determine whether or not it is a diamond" (CP 304). Frege also makes it quite clear that perceptual evidence will be required for this.

tions is extremely strict. It is, in fact, so strict that for practical purposes virtually none of our ordinary terms can be given appropriate definitions.

There are two distinct sorts of defects in our ordinary understanding of some concept-expressions which can prevent any definition that meets Frege's criteria from accurately capturing the content of the expression. These are ambiguity and vagueness. Since definitions of concept-expressions must guarantee that the concept determinately holds or does not hold for each object, a concept-expression that does not have universal application is ambiguous. An example is the arithmetical concept-expression 'prime'. Although primeness, as we understand it, determinately holds or does not hold of each integer, it is inapplicable to people. In order to obtain a definition of primeness which satisfies Frege's constraint, it will be necessary to supplement our everyday definitions of primeness by identifying the integers and stipulating, for each noninteger—person, idea, and so on—whether or not it is prime. Since, presuming it is possible to define the concept of integer, the choice of a particular stipulation of primeness or nonprimeness to other objects seems arbitrary, our everyday understanding of primeness seems ambiguous. The other sort of defect, vagueness, applies to everyday concepts with imprecise borders. An example is baldness, which, even when it is restricted to people, does not determinately hold or fail to hold in each case. Some people would be considered to be on the border between baldness and nonbaldness.

Since Frege's project has to do with arithmetic, the concepts whose imperfections he discusses tend to be concepts of arithmetic. These concepts are ambiguous because they are only taken to apply on limited ranges, but they generally are not vague. This is not to say that, on Frege's view, the main obstacle to providing definitions of arithmetical and scientific concepts is ambiguity. The task of defining a scientific concept is *not* primarily one of specifying its range and stipulating some arbitrary value the concept is to have outside its range. For, on Frege's view, some of the worst problems with nonsystematic sciences—the emergence of contradictions—result, not from the ambiguity of scientific concepts, but from their vagueness. Contradictions have appeared, Frege says, when scientists treat

> as a concept something that was not a concept in the logical sense because it lacked a sharp boundary. In the search for a boundary line, the contradictions, as they emerged, brought to the attention of the searchers that the assumed boundary was still uncertain or blurred, or that it was not the one they had been searching for. (CP 134)

Frege goes on to say that the "real driving force" behind the development of a scientific concept is "the perception of the blurred boundary" (CP 134).

The existence of ambiguous and vague scientific terms is partly a consequence of our general tolerance of ambiguous and vague terms. Everyday terms, of course, are almost invariably both ambiguous and vague. An example is the term 'bald'. Frege alludes to the defects in our use of this term when he says, "Bald people for example cannot be enumerated as long as the concept of baldness is not defined so precisely that for any individual there can be no doubt whether he falls under it" (COR 100). In order to get a sense of the scope of the defects involved, consider the work that would be involved in an attempt to give a Fregean definition of our everyday concept of baldness. In order to do so, not only would it be necessary to stipulate whether or not ideas, flavors, and cities are bald, it would also be necessary to say exactly how many hairs of exactly what thickness and in what density must be growing from a person's head before she does not fall under the concept 'bald'. Yet these defects do not prevent us from making use of the everyday understanding of the term 'bald'. Nor does Frege think that the precision and clarity he requires of scientific terms should be or could be required of our everyday terms (see, e.g., COR 115). Since baldness is not a mathematical concept, no definition of baldness will ever figure in a mathematical proof. Thus, on his own characterization, a definition of baldness would be completely worthless (FA 81).

At this point it may seem harmless to take Frege's understanding of definition to be nothing more nor less than an understanding of mathematical definition. But Frege's remarks concerning what is required of a definition cannot be read in isolation. Many of these remarks occur in the context of Frege's raising a more general worry about when a term is admissible for mathematical or scientific use. The conditions for admissibility of a concept-word are, more or less, the conditions that must be met by a definition of a concept-word. Frege says:

> But if we ask, under what conditions a concept is admissible in science.
> . . . The only requirement to be made of a concept is that it should have
> sharp boundaries; that is, that for every object it either falls under that
> concept or does not do so. (PW 179)[11]

[11]This requirement is the same as the requirement, mentioned in *Foundations*, for a concept-expression to be admissible for purposes of logic. Frege says: "All that can be demanded of a concept from the point of view of logic and with an eye to rigour of proof is only that the limits to its application should be sharp, that it should be determined, with regard to every object whether it falls under that concept or not" (FA 87). Frege then goes

It may seem odd to claim that the concept of baldness should be inadmissible for scientific use. However, any science in which use of this term might be necessary would be an empirical science, and it is possible to understand Frege's 'science' as 'mathematical science'.[12] Only two pages after the above passage, Frege writes, "But before we go on to proofs at all, we must have assured ourselves that the proper names and concept-words we employ are admissible" (PW 180). Since we never do go on to proofs in the empirical sciences, Frege's talk of admissibility may seem inapplicable to the concept of baldness.

One might be tempted to infer that Frege requires a special kind of precision of a mathematical (scientific) concept which would be inappropriate for empirical or everyday concepts. But Frege's comments about admissibility for scientific use cannot be viewed as confined to a limited mathematical domain. Frege's aim is to show that the laws of mathematics are derivable from the laws of logic—the laws of all thought. These laws must be everywhere applicable, not just applicable in a limited domain where extra precision is required. The sharp boundary condition is discussed in *Foundations* as what can be "demanded of a concept from the point of view of logic" (FA 87). The remarks about the necessity of sharp boundaries occur not only when he is talking about science but also when he talks about the demands of logic. He says,

---

on to state the corresponding conditions for object-words. He also says, in a later, unpublished, work: "In logic it must be presupposed that every proper name is meaningful; that is, that it serves its purpose of designating an object. . . . For instance 'the A' would be a name that was in this way inadmissible for scientific use if it were formed by means of the definite article from a name of a concept [*nomen appellativum*] under which either no object or more than one fell" (PW 179–180). The requirement that a term have sharp boundaries is, in fact, the characterization of admissibility for purposes of science (or of admissibility for logic—the most general science) which appears throughout Frege's work. See also CP 133, 134, 148; TWF 147; PW 122, 241.

[12]In fact, it would be wrong to understand Frege as talking about mathematical science only. The criteria that, for Frege, determine admissibility for science are identical with those that determine admissibility for logic. And he suggests, in the introduction to *Begriffsschrift*, that, after differential and integral calculus and geometry are translated into Begriffsschrift notation, it will be possible to proceed to "pure theory of motion and then to mechanics and physics" (BEG 7). He believes that the only important difference between mathematics and other sciences is that mathematics is more general. And he says, in *Foundations*, "Thought is in essentials the same everywhere; it is not true that there are different kinds of laws of thought to suit the different kinds of objects thought about" (FA iii). There is a great deal of evidence that Frege meant "admissibility for science" as admissibility for any science. However, it is not important to argue for this here. For, although it might seem that taking Frege to be making serious claims about empirical science will cause insuperable problems, in fact most of these problems will occur whether or not science is limited to mathematics. For the concept of number, as it is used by mathematicians, is every bit as problematic as the concept of baldness.

> If something fails to display a sharp boundary, it cannot be recognized in logic as a concept, just as something that is not extensionless cannot be recognized in geometry as a point, because otherwise it would be impossible to set up geometrical axioms. (CP 133)

and

> Contradictions have indeed appeared; but it is not as if they had been created by combining mutually contradictory characteristic marks in the definition; they have, rather, been created by treating as a concept something that was not a concept in the logical sense because it lacked a sharp boundary. (CP 134)

Admissibility for scientific use cannot be divorced from admissibility for logical use. But logic is supposed to be applicable to all thought. Thus, it seems that Frege has no room to allow what we might regard as our vague everyday concepts to play a role in our reasoning. In fact, in his later writings he explicitly denies that there are vague concepts. He says, in the second volume of *Basic Laws*,

> If we represent concepts in extension by areas on a plane, this is admittedly a picture that may be used only with caution, but here it can do us good service. To a concept without sharp boundary there would correspond an area that had not a sharp boundary-line all round, but in places just vaguely faded away into the background. This would not really be an area at all; and likewise a concept that that is not sharply defined is *wrongly termed a concept*. (TWF 139, my emphasis)

Elsewhere he says, of a concept that is not everywhere determinate, "In other words, what we have just called a concept would not be a genuine concept at all, since it would lack sharp boundaries" (TWF 148). He also calls concepts without sharp boundaries "inadmissible sham concepts" (TWF 145/GGA vol. 2, 74, section 62). And these remarks seem to apply to everyday concepts as well as scientific or mathematical concepts, for he asks, "Has the question 'Are we still Christians?' really got a sense, if it is indeterminate whom the predicate 'Christian' can truly be ascribed to, and who must be refused it?" (TWF 139). It may seem, then, that Frege's laws of thought, far from being everywhere applicable, in actual practice are *nowhere* applicable. Indeed, it may seem that, unless Frege can give some account of—or at least leave some room for the existence of—content in our imprecise everyday language, there is no content to be made precise for scientific purposes.

This may look to be a problem that results from the sort of interpretation I have been advancing. In particular, this may look to be a result of attributing to Frege a view of logic as the most general science. In fact, however, this problem arises in *Foundations*, regardless of how Frege's views about logic are interpreted.

Frege demands that 'number' either be defined or be recognized as indefinable (FA 5). It may be that, were he to recognize as convincing an argument that 'number' is indefinable, Frege would not need to deny the admissibility of this term for scientific use. Frege argues, however, that this term is definable. As a consequence he is committed to the view that this term was not acceptable for scientific use before his work. To see this, we have only to consider the significance of Frege's demand, in *Foundations*, that definitions determine sharp boundaries. Now it could be that Frege's definition is meant to do no more than spell out, explicitly, the mathematical content of the term 'number'.[13] But what is this content? What counts as acceptable evidence for claims about the content of this term? The evidence Frege seems to recognize as appropriate is mathematical practice, in particular the antecedent mathematical use of this term. Any claim to knowledge of what it is to be a number must be based on this. But evidence derived from our mathematical, or even mathematical and everyday, use of this term is not enough to determine the sharp boundaries Frege requires of a definition. For instance, Mill claims that numbers are properties of physical objects, while Frege claims that they are not. Yet it is highly implausible that Mill and Frege would have disagreed on any truth of arithmetic. No sophisticated knowledge of arithmetical truths or mathematical practice will suffice to refute either Mill's or Frege's view. In fact, more recent mathematical practice shows that mathematicians are certainly aware of the lack of sharp boundaries. For some purposes mathematicians today define the numbers as sets. On some of these set-theoretic definitions, $\{\{\{\phi\}\}\}$ is a number, and on others it is not. There is

---

[13]My use of the term 'content' here is deliberately vague. In *Foundations*, Frege had not yet drawn his celebrated *Sinn/Bedeutung* distinction. The words he uses, when he talks about terms and how they function in *Foundations*, are *Sinn* (sense), *Bedeutung* (reference, significance, or meaning), and *Inhalt* (content). It does not seem unreasonable to assume that, were I to have used some term other than 'content' here, some readers would be inclined to think I was using one of Frege's technical philosophical terms—a technical philosophical term that Frege had not yet formulated when he wrote *Foundations*. Although some of Frege's uses, in *Foundations*, of the terms I have mentioned contain most central features of his later technical understanding of these terms, other uses appear to be of terms for a vaguer, everyday notion. My use of 'content' is meant to mirror these uses.

no argument, among mathematicians, about whether or not {{{∅}}} is *really* a number. For mathematicians this question has no significance and no answer.

This highlights the strangeness of Frege's criteria of adequacy for definitions and admissibility for science. In spite of his objections, mathematicians had for some time used (and continue to use) 'one' and 'number' without any difficulty and without any definition of the sort Frege required. His formulation of definitions of these words did not change actual mathematical practice. Thus, it makes little difference for actual scientific practice whether or not a word is, in Frege's sense, acceptable for scientific use.

The point, of course, is that philosophical, not scientific, criteria determine, for Frege, what is acceptable for scientific use. His assumption is that every legitimate science must have a proper epistemological foundation. His purpose in making sure that 'one' and 'number' stand for acceptable concepts or objects is neither to show that some previously accepted mathematical results are actually false (true) nor to enable mathematicians to prove new results. Frege's real purpose is to make clear, from within his epistemological perspective, how it is that we can have knowledge of truths of arithmetic. As he argues in the first chapter of *Foundations*, our ordinary use of 'one' and 'number' does not make this clear. However, it is not obvious that giving criteria for determining whether or not something is a number (or whether or not something is the number one) should enable us to see the ultimate grounds of the justification of claims about numbers. For there seems to be a tension between Frege's notion of definition and the role definitions are to play in establishing the ultimate grounds of our knowledge of arithmetic. Frege's criteria of adequacy for definitions can be satisfied only by definitions that are arbitrary stipulations. But how can an arbitrary stipulation tell us anything about the grounds for our knowledge of arithmetic? It will be useful to begin by considering what is involved in showing that a term is, in Frege's sense, acceptable for scientific use.

First of all, it is important to note that definition is not the only means available. Frege says, in *Foundations*, that the project of finding the ultimate grounds of the truths of arithmetic requires that 'number' either be defined or be recognized as indefinable (FA 5). In *Foundations*, of course, Frege argues that the terms he is discussing are in fact definable. Hence, there is little discussion in that work of indefinable terms. Of the indefinable (or, at least, undefined) term 'extension of a concept', which Frege uses in his definitions, he says only, "I assume that it is known what the extension of a concept is" (FA 80) and "In this definition the sense of the expression 'extension of a concept' is assumed to be known" (FA 117). In some of his later writings, he is more explicit about the treatment of indefinable terms.[14] He says, "So long as it is not completely defined, or known in some other way, what a word or symbol stands for, it may not be used in an exact science" (TWF 140). What other ways are there? After indicating, in "On Concept and Object," that not all terms are definable, Frege says that in order to explain what is meant by a term that is to stand for something logically simple, that is, indefinable, "I must then be satisfied with indicating what I intend by means of hints" (BLA 32). Hints, of course, are less satisfactory than definitions. Hints cannot be used in proofs or scientific investigations and, if their use is to be successful, a certain amount of cooperative understanding is required. Consequently, hints are only acceptable alternatives to definitions if the term in question is a term for something logically simple and cannot be defined. All other terms must be defined and, moreover, defined from primitive terms ("our endeavour must be to trace our way back to what is logically simple," BLA 32).[15]

The relation between Frege's epistemological concerns and this understanding of what is required to make clear what a term stands for stems from the epistemological features of primitive elements. For Frege assumes that the source of our knowledge about primitive (logically simple) elements will be obvious.[16] Thus, there are epistemological reasons for defining a term even when there is no doubt about what it stands for. Frege says:

> The real importance of a definition lies in its logical construction out of primitive elements. And for that reason we should not do without it, not

[14]Actually, Frege does argue, in *Basic Laws* (sections 146–147; see GGA vol. 2, 147–149), that a use of the notion of extension of a concept is implicit in the statements of many mathematicians. He also offers some informative discussion of the meaning of this notion. I will discuss these sections, as well as the related sections of *Foundations* (64–69), later when I look more closely at the notion of primitiveness.

[15]This notion of something logically simple is distinct from that of a simple sign or of what is designated by a simple sign. A simple sign is a sign that is not composed of signs that have other uses. Such a sign can be introduced as the abbreviation of a complex expression (see, e.g., BLA 90, PW 208). Indeed, when a definition is used to stipulate the meaning of a term, the sign must be simple. Thus, a simple sign will often designate something logically complex.

[16]For instance, the basis of Frege's argument in *Foundations* that 'number' and 'one' are not primitive is that we do not immediately recognize the source of our knowledge about numbers.

even in a case like this. The insight it permits into the logical structure . . .
is a condition for insight into the logical linkage of truths. (CP 302)

This insight into the logical linkage of truths, along with our access to
the source of the justification of claims about primitive elements, will
help us to determine the grounds for our justification of such things as
the statements of arithmetic. For, if all nonprimitive terms in a sentence
are defined from logically simple terms, then the statement expressed
by the original sentence can also be expressed by a sentence containing
only primitive terms. Since the grounds of our knowledge about primitive elements will be obvious, the new expression of this statement will
make the grounds of our knowledge obvious. Hence, to find the ultimate grounds of our knowledge of some statement, it will suffice to
determine the primitive elements involved, or to replace all nonprimitive terms in the sentence by which it is expressed with defining expressions all of whose terms are primitive. The implicit picture is that any
science is based on a small number of logically primitive elements and,
consequently, that its truths can be expressed by introducing a small
number of primitive terms from which all other terms can be defined (or
into terms of which all other terms are analyzable). Frege says:

> To be sure, that on which we base our definitions may itself have been
> defined previously; however, when we retrace our steps further, we shall
> always come to something which, being a simple, is indefinable, and
> must be admitted incapable of further analysis. And the properties be-
> longing to these ultimate building blocks of a discipline contain, as it were
> in a nutshell, its whole contents. . . . Now it is clear that the boundaries of
> a discipline are determined by the nature of its ultimate building blocks.
> (CP 113)

Thus, when Frege asks for a definition of some term of some science, he
is asking not only what the term stands for, but also what the fundamental building blocks of that science are. The purpose of finding these
is to determine its ultimate grounds.

At this point it is possible to begin to address the issue of how Frege
can regard definitions as arbitrary abbreviations and also as contributing to this sort of foundational analysis. He says:

> My opinion is this: We must admit logically primitive elements that are
> indefinable. Even here there seems to be a need to make sure that we
> designate the same thing by the same sign (word). Once the investigators
> have come to an understanding about the primitive elements and their
> designations, agreement about what is logically composite is easily

reached by means of definition. Since definitions are not possible for primitive elements, something else must enter in. I call it explication. It is this, therefore, that serves the purpose of mutual understanding among investigators as well as of the communications of the science to others. (CP 300)

This appears to fit in neither with the view of definitions as arbitrary abbreviations nor with the view of definitions as consisting of analyses of the terms that are to be defined. For the view expressed in this passage seems to be a view of defining as a part of the building of a science from the foundations up. But it is not clear how a term can be simultaneously analyzed and constructed out of foundations. It would seem either that the term is already (implicitly or explicitly) constructed out of primitive terms, in which case the analysis consists of merely elucidating what is already there (and hence the term is not being constructed), or that the term is being constructed initially from primitive terms, in which case there is no antecedent content to be analyzed. *Neither* of these stories seems to fit in with Frege's project of defining the number one. Frege cannot be creating a new science by constructing a definition of 'one' from primitive terms, for there already is a science of arithmetic. But Frege also cannot be merely elucidating the content of 'one' by constructing, in primitive terms, a description that picks out what 'one' already stands for. For the truths of arithmetic do not determine a particular object for which 'one' must stand. There appears to be a serious tension in Frege's view about the role of definition in his overall project.

There are only hints of a response to this tension in Frege's early work. But a more complete response does exist in a later, unpublished, article, "Logic in Mathematics." Before I discuss this response, it will be useful to make a few observations about the discussions of definition in the preceding pages.

Foundations was written before the 1892 paper in which Frege introduced his celebrated *Sinn/Bedeutung* distinction. The term *Sinn* is typically translated by the English term 'sense', but there has been a great deal of disagreement about the translation of *Bedeutung*. Among the English words used to translate it are 'denotation', 'reference', 'meaning', and 'significance'. Except for 'significance', all of these words carry a great deal of late twentieth-century philosophical baggage—baggage that, on my own interpretation, Frege's term *Bedeutung* does not carry. In order to avoid begging the questions I am about to raise, I have tried to avoid using any of these terms and, instead of writing about the pre-Fregean reference or meaning of the word 'one', I have written of its pre-Fregean content. My use of the word 'content' is deliberately vague. This vagueness is justified, I think, because, first, although it is undeniable that both numerals and the sentences that are taken to express truths of arithmetic had some sort of content before Frege's work, it is not clear exactly what, on Frege's view, this content could be and, second, whatever this content is, the tension described above seems worrying. For sentences, my use of 'content' is meant to correspond to Frege's early use of 'judgeable content' (*beurtheilbarer Inhalt*—roughly, the content of an assertoric sentence) which is later divided, by his *Sinn/Bedeutung* distinction, into thought and truth value.

Once Frege had drawn his *Sinn/Bedeutung* distinction, the terms *Sinn* and *Bedeutung* appear to have become technical philosophical terms. "Logic in Mathematics" was written long after 1892. Thus, the occurrence of these terms in the passages in "Logic in Mathematics" which, I will argue, provide Frege a response to the difficulties discussed above must be regarded as an occurrence of technical philosophical terms. But once these occurrences are regarded as occurrences of technical philosophical terms, my claim that "Logic in Mathematics" provides a response to a tension in *Foundations* seems less plausible. Part of my defense of this interpretation of "Logic in Mathematics" will come from my earlier argument that Frege's views on definition did not change. This argument was based both on Frege's claims that his views on definition did not change and on the similarity, over the course of his career, of his characterizations of definition. And definition, on these characterizations, is intimately connected with *Bedeutung*. He says, in *Foundations*, that a definition lays down the meaning (*setzt die Bedeutung*) of a symbol (FA 78) and, in "Logic in Mathematics," that a meaning is bestowed upon a sign by a definition (*durch eine Definition einem Zeichen eine Bedeutung gegeben worden ist*, PW 208/NS 224), and that this bestowal of meaning on a term is the role of definition (PW 244). But if *Bedeutung* is a technical philosophical term in "Logic in Mathematics" which Frege introduced after *Foundations* was written, how can these quotations be understood in the same way?

My strategy for answering this question is to begin with a brief summary of Frege's exposition of his *Sinn/Bedeutung* distinction and then to argue that the technical philosophical understanding of the term *Bedeutung* is actually implicit in the use, in *Foundations*, of this term. In fact, it seems that Frege's use of *Bedeutung* in *Foundations* is not really vaguer than his post-1892 use but, rather, that Frege sometimes uses this term in an everyday sense and other times in what appears to be the technical sense introduced later. I will argue that, in particular, there is one context—his discussions of definition—in which the *Foundations* use of *Bedeutung* is indistinguishable from his later uses of this term. For the effect of the introduction of the sharp boundary constraint on definitions is to give his *Foundations* claims that definitions fix the meaning (*Bedeutung*) of a sign precisely the content these claims have in his later writings. Once I have argued that there is an appropriate connection between Frege's claims about definition and meaning (*Bedeutung*) in *Foundations* and in "Logic in Mathematics," it will be possible to find a response in "Logic in Mathematics" to the tension in Frege's view about the role of definition in his project. In the discussion that follows, I will be treating 'Bedeutung' as an English term.

Frege's introduction of the distinction in "On Sense and Meaning" ("Über Sinn und Bedeutung") is disarmingly simple and appears to have no obvious relation to the problems I have been discussing. If identity is simply a relation that an object bears to itself and nothing else, what, he asks, can account for the difference in the information conveyed by the claims that a = a and a = b when both are true? Frege's answer is to distinguish the object for which a proper name stands (its Bedeutung) from the sense (*Sinn*) of the proper name. The sense of a proper name is its mode of presentation of the object that is its Bedeutung. The expressions 'evening star' and 'morning star' via distinct modes of presentation present the same thing. These expressions have the same Bedeutung but different sense. Sentences, too, have both sense and Bedeutung. Frege says that a sentence contains a thought. This thought is the cognitive content of, or information conveyed by, the sentence. It is composed of the senses of the parts of the sentence. Frege then argues at some length that the Bedeutung of a sentence must be its truth value. Although Frege does not discuss concept-expressions in this paper, some later unpublished remarks make it clear that the sense/Bedeutung distinction applies to these expressions as well.[17]

At first glance, Frege's "On Sense and Meaning" may seem to contain nothing more nor less than an articulation of some easily recognizable features of the workings of language. It does not take much time,

---

[17]See "Comments on Sense and Meaning," PW 118–125. The view that a concept has Bedeutung (and that the Bedeutung is distinct from its extension) does not appear solely in Frege's unpublished writings. See also, for instance, CP 205, 227, 303, 307.

however, for this impression to fade. While it may seem that the relation between a proper name and its Bedeutung is simply that of a name to its bearer, the relation between a sentence and its truth-value hardly seems to be the same relation. Indeed, most of "On Sense and Meaning" is devoted to an attempt to make his odd claim that the Bedeutung of a sentence is its truth-value seem plausible. Reflection on this claim, in particular, makes Frege's understanding of the term 'Bedeutung' look more like that of a technical philosophical term than of an everyday term.[18] Consideration of some of his later remarks about Bedeutung, in fact, begins to cast a rather different light even on Frege's apparently unexceptionable discussion of proper names in "On Sense and Meaning."

Although his concern in the discussions of "On Sense and Meaning" looks to be a concern with natural language, in his later papers his remarks about language or Bedeutung almost invariably occur in the context of discussions of logic or science. These remarks typically arise in the context of a discussion of what is required of a term if it it is to be admissible for purposes of science. Frege says that what is required of a term if is to be admissible for scientific purposes is precisely that it have Bedeutung.[19] For instance,

> In scientific use, a proper name has the purpose of designating an object; and in the case this purpose is achieved, this object is the meaning of the proper name. The same thing holds for concept-signs, relation-signs, and

[18]According to Michael Dummett, Frege's claim that an assertoric sentence has as its referent a truth-value is a serious blunder. See *Frege: Philosophy of Language*, 180–186 (FPL). The view of truth-values as objects and sentences as names of truth-values requires Frege to provide senses for sentences in which a concept is ascribed to a truth-value—for sentences like '(5 = 2 + 4) is green'. The problem with this, on Dummett's view, is that Frege's aim was to give an account of the workings of language and that the logical functions of sentences are different from those of the other expressions that Frege would identify as proper names. Because the aims of Frege's project, as I interpret them, are somewhat different from the aims Dummett attributes to Frege, I am unconvinced by Dummett's arguments that Frege needed to provide a theory of the workings of language. As I will argue, in chapter 6, it is a consequence of Frege's aims, as I interpret them, that there cannot be such a theory. In fact, my reasons are not dissimilar from some of Dummett's reasons for arguing that, on Frege's view, there cannot be a complete theory of meaning or semantics for any natural language (FPL 106–107). In "Frege on Truth," Tyler Burge takes issue with Dummett's objection. Burge's reasons also result from an interpretation on which Frege's concern is not with ordinary language.

[19]Gregory Currie has also suggested that Frege's interest in the sense/reference, or sense/meaning, distinction has to do with his views about science and not with his views about ordinary language or meaning. See *Frege: An Introduction to His Philosophy*, 102–103.

function-signs. They designate concepts, relations, and functions respectively, and what they designate is then their meaning. (CP 298)

And

In science the purpose of a proper name is to designate an object determinately. (PW 178)

This does not seem unreasonable and it does not seem to divorce admissibility for scientific purposes from what is required of terms used in natural language. After all, it seems to amount simply to saying that, for purposes of scientific research, we want to be talking about things in the real world, not some phantasms of the imagination. Such a requirement may seem to fit everyday use of natural language equally well. And, on this interpretation, it is not surprising that he contrasts what is required of terms in scientific enterprises—having Bedeutung—with what is required for purposes of myth and fiction. But the coincidence of this notion of Bedeutung with our everyday notion of what a term stands for does not survive a more detailed consideration of Frege's view.

It is important, first of all, to note that in Frege's later writings he is as insistent about the necessity, for scientific purposes, that concept-expressions have Bedeutung as he is that object-expressions have Bedeutung. The significance of the mention, in the above quote from "Foundations of Geometry II" (CP 298), of the requirement that concept-signs designate concepts becomes clearer when we consider Frege's notion of concept. As we have seen earlier, a concept must have a sharp boundary—it must be determinately true or false of each object. He says, "It must be determinate for every object whether it falls under a concept or not; a concept word which does not meet this requirement is meaningless [bedeutungslos]" (PW 122/NS 133).[20] He also says:

---

[20]This is not an isolated claim. Similar remarks can be found in "Function and Concept," CP 148; "On Sense and Meaning," CP 169; "On the Foundations of Geometry II," CP 303; and Basic Laws section 64 (TWF 148). Frege suggests, in Foundations, that a symbol can signify something only if we can determine, for any object, whether it is the same as the object signified (FA 73), and that it is only possible to say what 1 + i means if it can be determined whether or not 1 + i is the idea of an apple (FA 113). He suggests, in an unpublished article, "Über Schoenflies: Die logischen Paradoxien der Mengenlehre," that a proper name of the form 'the A' is meaningful only if exactly one object falls under the concept.

Thus the requirement we have here set up—that every first-level function of two arguments must have an object as its value for any one object as its first argument and any other objects as its second—is a consequence of the requirement that concepts must have sharp boundaries and that we may not tolerate expressions which seem by their structure to mean a concept [*einen Begriff zu bedeuten*] but only create an illusion of so doing, just as we may not admit proper names that do not actually designate an object. (TWF 148/GGA vol. 2, 77, section 64)

There is an obvious connection between these remarks and the *Foundations* remarks about what is required of definitions of one and the concept number. For in these remarks Frege says that what must be determinate for a term to have Bedeutung is that it has a sharp boundary. Thus, Frege's sharp boundary constraint on definition, in *Foundations*, guarantees that a defined term has, in Frege's later sense, Bedeutung. In this context, there is no difference between Frege's claim in *Foundations* that a definition lays down the meaning (Bedeutung) of a term (FA 78) and his similar claims in later writings, among which is the "Logic in Mathematics" comment that a definition bestows a meaning (Bedeutung) on a term (PW 208).

My reason for introducing 'Bedeutung' as an English term should now be evident. For, once the connection has been made between Frege's notion of Bedeutung, definition, and admissibility for purposes of science, Bedeutung looks very different both from current technical philosophical notions of reference and meaning and from the related everyday notions. Although we take our everyday terms to have reference and meaning, it is a consequence of Frege's view that few of our everyday terms can have Bedeutung.[21] Consider, again, the everyday term 'bald'. In order for this term to have Bedeutung, in Frege's sense, it must be determinate precisely which objects are and which are not bald. But there are a great many objects that we would not count as bald or as not-bald either because they seem to be out of the range of the concept (e.g., ideas) or because they seem to be borderline cases. The former problem is not obviously insuperable. After all, provided we could explicitly define what is now taken to be the range of baldness, we could stipulate that everything outside this range fails to be bald. It is not at all

[21]Jean van Heijenoort and Tyler Burge have also suggested that, due to their vagueness, words in ordinary language have no Bedeutung on Frege's view. See van Heijenoort, "Frege and Vagueness"; Burge, "Frege on Extensions of Concepts from 1884 to 1903"; and Burge, "Frege on Truth."

clear that it would be wrong to say that the resulting concept is what was always meant by baldness. It is less easy, however, to see what should be said about the latter problem. Our everyday notion of baldness is vague. Frege's characterization, in *Basic Laws*, of inadmissible sham concepts (TWF 145) is obviously applicable. If our everyday term 'bald' corresponds only to an inadmissible sham concept, it cannot, on Frege's view, have Bedeutung.

It may seem, however, that the ascription of the everyday notion of reference to everyday concept-expressions is sufficiently tenuous that the above considerations do not really militate against an interpretation of Fregean Bedeutung as reference. The upshot of these consequences for everyday concept-expressions might simply be taken to be that more precision is required of concept-expressions in scientific contexts than in everyday contexts. The notions of having Bedeutung and having reference look to be the same for proper names. But Frege's treatment of concept-expressions has consequences for object-expressions (proper names) as well.

Definite descriptions, which are used both in scientific and in everyday contexts, are among the expressions that Frege counts as proper names. Although Frege does not give any universal criteria for identifying the circumstances under which a proper name designates an object, he does give criteria for identifying the circumstances under which these particular proper names designate objects. Frege says that unless precisely one object falls under a concept, A, then 'the A' is inadmissible for scientific use, and "if A is the concept of a right-angled equilateral pentagon, 'this A' is an inadmissible—meaningless—proper name [*ein bedeutungsloser unzulässiger Eigenname*]" (PW 180/NS 195). In this particular circumstance, Frege is not considering the possibility that 'A' is not an admissible concept-expression but rather the possibility that the concept in question is not true of anything. On the other hand, if there is no admissible concept, then we cannot say that precisely one object falls under it. Frege says, in *Foundations*, "the use of the definite article cannot be justified until we have first proved two propositions" (FA 89). These two propositions are that some object falls under the concept and that only one does (FA 87–88). Such proofs can only be obtained if we have a real concept, something that is determinately true or false of each object.

These criteria for the admissibility of proper names that are definite descriptions preclude many proper names of natural language from having Bedeutung, in Frege's sense, although we regard them as designating objects. The problem comes from our use, in natural language, of inadmissible concept-expressions. In many everyday situations the vagueness of our everyday understanding of baldness, maleness, and location in a corner, for instance, will provide no bar to our confidence that the expression 'the bald man in the corner' applies to precisely one object.[22] In such a case, it would seem, we would want to say that the expression designates that object. Can Frege agree with the claim that, in certain circumstances, 'the bald man in the corner' has Bedeutung? Or, more specifically, must such agreement conflict with anything Frege says about definite descriptions?

There is no question that such agreement would conflict with the passages on definite description mentioned above. Nor does Frege's discussion of definite descriptions in "On Sense and Meaning" (CP 168–170) suggest any way out. This discussion is introduced when Frege considers the following example:

> Whoever discovered the elliptic form of the planetary orbits died in misery. (CP 168).

He says that whether or not the subordinate clause in this sentence has a Bedeutung depends on the truth of the sentence:

> There was someone who discovered the elliptic form of the planetary orbits. (CP 169).

The truth of this sentence will, of course, depend on all its constituents' having Bedeutung. In particular, then, the concept-expressions involved must designate concepts.

Does Frege's view allow any room for 'the bald man in the corner' to designate an object (to have Bedeutung) if 'bald man in the corner' fails to designate a concept? I think there are some available strategies, but

---

[22]I am ignoring sceptical examples here. If it turns out that there is exactly one person in the corner and that person is a woman who, because of her baldness and masculine attire, is presumed to be a man, then it follows that 'the bald man in the corner' does not refer to anyone. The point has nothing to do with our capacity to determine whether or not certain object-expressions do refer to objects, and the point has nothing to do with solving puzzles about reference. Rather, the point is that it is not obvious that Frege's views about Bedeutung conflict with our views about the reference of everyday expressions. If this can be established, it may then be appropriate to turn to the question of whether or not what Frege says about Bedeutung can be developed into an acceptable theory of reference. Since I intend to argue that this lack of conflict cannot be established, the question about acceptable theories of reference is irrelevant.

none that accords with our everyday view of this expression.[23] One such strategy would be to make use of another definite description that is admissible and that picks out the appropriate object. If so, one could, by definition, stipulate that 'the bald man in the corner' is to designate that object. But it is not clear that we would have gained anything through such a maneuver. For the point of any use of the expression 'the bald man in the corner' has to do with the sense of our everyday (inadmissible) notions of baldness, maleness, and being in a corner. We use this sort of description because we think that these component parts will enable the hearer to pick out the intended object. If the condition under which the expression 'the bald man in the corner' can be used is that it be introduced via a stipulative definition, using different concept-expressions, then there would be no point to the use of this expression rather than, for instance, the defining expression.

The only other strategy that seems applicable is to regard 'the bald man in the corner' as a primitive term. It is not obvious that there is any reason for prohibiting its introduction as a primitive term. And we might be inclined to think that under appropriate circumstances this expression, as well as any other successful everyday proper name, could be used for scientific purposes as a primitive term. But, as with the previous strategy, this appears to prohibit the decomposition of 'the bald man in the corner' into its parts. There is also another worry about this strategy—it is not clear that it is appropriate, on Frege's view, for there to be primitive object-expressions. I will discuss this issue later. In this later discussion, I will also consider the sort of proper names which seem to have the most claim to primitive status—real proper names, for example, 'Gottlob Frege'.

Throughout the discussion of the Bedeutung of everyday proper names, I have emphasized the fact that, except for some of the remarks in "On Sense and Meaning," Frege's discussions of the notion of Bedeutung (or, for an object-name, of its designating an object) occur in the context of a discussion of scientific terms. Indeed, he makes such remarks as "How things may be in ordinary language is of no concern of

[23]There is another alternative, although I do not discuss it below because it is not relevant. In one of Frege's unpublished works he asks whether the proper name "the set of all sets which do not contain themselves as elements" is meaningful (*bedeutungsvoll*, PW 181/NS 196), and he concludes that it is not a definite description. A similar discussion appears in *Basic Laws* with respect to his introduction of expressions for the extension of concepts (TWF 159–160). His reasoning, however, does not apply to 'the bald man in the corner', which really is understood as a definition description. And his solution, in the former case—that a logical law is necessary—seems entirely inappropriate.

ours here" (PW 178–179). Even some of the passages in "On Sense and Meaning" seem to have been written for the purpose of enlightening us about admissibility for scientific purposes. The discussion of the sentence "Whoever discovered the elliptic form of the planetary orbits died in misery" is designed, at least in part, to exhibit a sort of locution which should be prohibited in scientific contexts.[24]

What are we to make of the apparent assumption in "On Sense and Meaning" that everyday terms *do* have Bedeutung? Given Frege's strict adherence, throughout his writings, to the sharp boundary criterion for admissibility of scientific terms, it might seem reasonable to suppose that Frege would recognize some looser criteria for the attribution of Bedeutung to everyday terms. There is no indication of this in Frege's writings, however. Frege never distinguishes scientific-Bedeutung from everyday-Bedeutung. On the other hand, he does talk about Bedeutung and everyday language, in a letter to Peano dated 29 September 1896, where he explicitly denies that a term must have Bedeutung if it is to be used successfully in everyday language. He says:

> Thus a sign for a concept whose content does not satisfy this requirement is to be regarded as meaningless [*bedeutungslos*] from the logical point of view. It can be objected that such words are used thousands of times in the language of life. Yes; but our vernacular languages are also not made for conducting proofs. . . . The task of our vernacular languages is essentially fulfilled if people engaged in communication with one another connect the same thought or approximately the same thought, with the same proposition. For this it is not at all necessary that the individual

[24]It is interesting to note that even Dummett, who is perhaps the most vehement defender of the view that Frege was a philosopher of language, and who says, "The analysis of language which Frege undertook involved an analysis of the *working* of language" (FPL 81), suggests that the project of giving an analysis of language was a necessary part of the *Begriffsschrift* project of constructing a symbolic language adequate for mathematical theory. Nor does Dummett argue that Frege intended to give an analysis of natural language. Dummett also notes that, on Frege's view, there are well-formed sentences of natural language which do not have a truth-value. Dummett comments, "It is of the greatest importance, for the understanding of Frege, to grasp that, while this was, for him, the correct account of how matters stand with regard to natural language, it was a totally unsatisfactory state of affairs, revealing a defect of natural language, which must be remedied in any properly constructed language such as Frege's own symbolic language. . . . He thought, that is, that it is impossible to give any coherent account of the functioning of a language in which it is possible to construct well-formed sentences which lack a truth-value" (FPL 166–167). Thus, although Dummett would certainly disagree with my characterization of the *Sinn/Bedeutung* distinction, this is a consequence neither of my interpretation of the introduction of the *Sinn/Bedeutung* distinction as a contribution to Frege's inquiry into the foundations of arithmetic nor of my view that Frege's *Sinn/Bedeutung* distinction does not apply to natural language.

words should have a sense and meaning of their own, provided only that the whole proposition has a sense. (COR 115/WB 183)

In this passage, Frege qualifies his attribution of meaninglessness to concept-expressions without sharp boundaries with the phrase "from the logical point of view." Thus, one might suspect that there is another notion of meaning (Bedeutung) appropriate for our imprecise everyday language. Had Frege believed that there was a serious alternative notion of Bedeutung for everyday language, this would be a natural place for him to say so. But he says nothing. Mostly Frege simply avoids talking about Bedeutung and ordinary language. In general, when he contrasts terms with scientific-Bedeutung from other terms that are used but have no Bedeutung, the latter are generally taken not to be terms used in everyday communication but, rather, terms appearing in fiction or poetry.[25]

There seem to be no grounds on which to attribute to Frege the view that the conditions everyday terms must satisfy in order to have Bedeutung are distinct from those that must be satisfied by scientific terms. It seems that Bedeutung and the criteria for a determining whether a term has Bedeutung cannot vary—that there is no everyday-Bedeutung versus a scientific-Bedeutung. We are left with a question about the apparent deception in "On Sense and Meaning." If few everyday terms actually have Bedeutung, in Frege's sense, why should Frege have introduced this notion in a paper in which he writes as if everyday terms do have Bedeutung? It will be easier to answer this question if we consider the role Frege's sense/Bedeutung distinction plays in resolving tensions in his project of defining the numbers.

Frege wants definitions of the terms 'one' and 'number'. These definitions, along with certain proofs, are supposed to allow us to see what the ultimate ground of arithmetical truth is. That is, Frege means to be enlightening us about terms that are already in use in a science that is already practiced. It does not seem unreasonable to expect an analysis of *our* terms and *our* science. And, at least at some points, Frege seems to be providing just that. For instance, when he claims in *Foundations* that assertions of number are assertions about concepts (FA 59), he seems to be saying something genuinely enlightening about our understanding of the positive integers. The observation that, for instance, the claim that there are three letters in the word 'red' can be understood as

<hr/>

[25]See, for instance, PW 122, 186, 198, 232; CP 226, 241, 329; COR 63, 80, 152, 165.

$(\exists x)(\exists y)(\exists z)(x, y,$ and z are letters in the word 'red', $x, y,$ and z are pairwise distinct, and every letter in the word 'red' is either x or y or z)

tells us something about *our* understanding of the number three without adding anything other than a new notation. Yet our pre-Fregean understanding of the positive integers and the concept number does not determine sharp boundaries. Hence, these terms do not have a Bedeutung, in Frege's *Foundations* sense, and Frege's definitions cannot simply be expressions, all of whose terms are primitive, which have the same Bedeutung as the terms 'one' and 'number'. That is, Frege's definitions cannot simply be analyses of these terms, they must go beyond our pre-Fregean understanding of arithmetic. In what sense, then, does the science built up from primitive terms via Frege's definitions of these terms deserve to be called 'arithmetic'? In what sense does information about the ultimate ground of this science tell us something about the ultimate ground of arithmetic?

To see how "On Sense and Meaning" can provide Frege with a response to this tension, it may help to look at it in another way. The pre-Fregean use or understanding of our arithmetical symbols does not determine a Bedeutung for them. Furthermore, before "On Sense and Meaning," Frege uses the terms 'content' (*Inhalt*), 'sense' (*Sinn*), and 'meaning' (*Bedeutung*) almost interchangeably.[26] And having sense (or content or meaning) seems to constitute the difference between linguistic terms and completely empty marks or inscriptions. But given this, it is difficult not to infer, from his criticisms of the views about numbers of others, his explication of his very strict criteria for a term's admissiblity for scientific purposes, and his claim that the numeral '1'—as it is used before his definition—does not meet these criteria, that Frege's philosophical views lead him to reject pre-Fregean arithmetic as a game played with empty inscriptions.

Yet, as his early discussions of formalism show, Frege thinks it is

---

[26]For instance, Frege's use of *Inhalt* in the *Begriffsschrift* introduction of the identity sign (section 8) corresponds to his later use of *Bedeutung*. To see this, one only need compare this section with its description in the opening passages of "On Sense and Meaning." Also, Frege's use, in sections 97–103, of *Sinn*, as Frege later says in a letter to Husserl (COR 63), corresponds to his later use of *Bedeutung*. This is not to say that Frege's use of these terms in his early works can always be interchanged, but a survey of all the uses to which these terms are put would not be especially informative. My point here is that these terms do not play distinct roles until after "On Sense and Meaning" and, consequently, that Frege's criticisms of the scientific uses of '1' and 'number' appear to leave no legitimate role for these terms to play.

illegitimate to take the numerals and other signs of arithmetic as empty symbols. As he says in his *Basic Laws* discussion of Thomae, what distinguishes our arithmetic from a purely formal game is its applications. "Now it is the applicability alone which elevates arithmetic from a game to the rank of a science. So applicability necessarily belongs to it. Is it good, then, to exclude from arithmetic what it needs in order to be a science?" (TWF 167). Frege's *Foundations* criteria for acceptable definitions of the signs of arithmetic are designed to capture the applications of arithmetic—for instance, the above analysis of the content of "There are three letters in the word 'red.'" But the view that there is some content to the pre-Fregean use of the symbols of arithmetic which determines the constraints Frege's definitions must satisfy can make no sense as long as all Frege's terms for content (*Inhalt*, *Bedeutung*, and *Sinn*) are linked to his criteria for a term's admissibility. What is needed is some sort of content which falls short of these criteria, which can be ascribed to the pre-Fregean understanding of the symbols of arithmetic, and which Frege's definitions can be seen as making more precise. This content is sense. The sense of an assertoric sentence is a thought. Thus, Frege is able to say, in *Basic Laws*,

> For an arithmetic with no thought as its content will also be without possibility of application. Why can no application be made of a configuration of chess pieces? Obviously, because it expresses no thought. If it did so and every chess move conforming to the rules corresponded to a transition from one thought to another, applications of chess would also be conceivable. Why can arithmetical equations be applied? Only because they express thoughts. How could we possibly apply an equation which expressed nothing. . . ? (TWF 167)

The formulation of the notion of sense now allows Frege to formulate a response to the tension concerning analysis.

The tension arises from the observation that, given the inadequacy of the pre-Fregean understanding of the content of the terms of arithmetic, no admissible definitions can truly constitute analyses of the pre-Fregean notions. But, given that Frege's definitions do not provide analyses of the pre-Fregean arithmetical concepts, it is not clear how a science founded on these definitions can provide ultimate grounds for our arithmetic. An explanation can be found in a 1914 unpublished paper, "Logic in Mathematics." Frege says:

> Now we shall have to consider the difficulty we come up against in giving a logical analysis when it is problematic whether this analysis is correct.

Let us assume that A is the long established sign (expression) whose sense we have attempted to analyse logically by constructing a complex expression that gives the analysis. Since we are not certain whether the analysis is successful, we are not prepared to present the complex expression as one which can be replaced by the simple sign A. If it is our intention to put forward a definition proper, we are not entitled to choose the sign A, which already has a sense, but we must choose a fresh sign B, say, which has the sense of the complex expression only in virtue of the definition. The question now is whether A and B have the same sense. But we can bypass this question altogether if we are constructing a new system from the bottom up; in that case we shall make no further use of the sign A—we shall only use B. . . . If we have managed in this way to construct a system for mathematics without any need for the sign A, we can leave the matter there; there is no need at all to answer the question concerning the sense in which—whatever it may be—this sign had been used earlier. . . . However, it may be felt expedient to use sign A instead of sign B. But if we do this, we must treat it as an entirely new sign which had no sense prior to the definition. (PW 210–211)

This, of course, is a description of Frege's project of defining the number one. Frege *does* regard himself as building a science from the bottom up. But Frege is not building an arbitrary science from the bottom up—he is building a science that can replace arithmetic. A science that can replace arithmetic, of course, must be a science that has all the applications of arithmetic.

Before I explain how this description of Frege's project provides him with a response to the apparent tensions in his notion of definition, a brief digression will be necessary. In *Foundations*, the arithmetic with which Frege seems to be concerned is the arithmetic of numbers that Frege characterizes as being ascribable to concepts—that is, the positive integers and 0. In fact, however, the arithmetic that Frege wants to replace is arithmetic for real numbers, numbers that, on Frege's characterization, are not ascribed to concepts but are used in measurement. Frege gives definitions of the first sort of number in volume 1 of *Basic Laws* and uses these, in volume 2, to define the real numbers. In fact, numbers of the first sort, as Frege defines them, are *not* real numbers. In the interest of simplicity, I have described Frege's project below as if his aim were to replace, not arithmetic for real numbers, but arithmetic for positive integers.

The strategy for constructing such a science is as follows. First, there must be a means for translating sentences of the new science into sentences of the old and sentences of the old into sentences of the new. Sentences of the new science will be Begriffsschrift expressions supplemented by certain defined terms, while sentences of the old science will be our familiar sentences of arithmetic, for example, '1 + 1 = 2'. Second, there must be a serious sense in which these are translations. That is,

1. the translation, into Begriffsschrift, of what has previously been regarded by mathematicians as expressing a truth of arithmetic must express a provable truth of Begriffsschrift

and

2. the translation must enable what has previously been regarded as a correct proof of some truth of arithmetic to be replaced by a correct (i.e., gapless Begriffsschrift) proof in Frege's constructed science.

How is such a translation to be constructed?

It is important to note, first of all, that Frege believed some of the structure of everyday proof is already expressible in his Begriffsschrift. According to Frege, all but the most basic truths of arithmetic are provable, by means of general logical laws, from these basic truths (e.g., the associative law for addition) and definitions of the numbers from their predecessors. That is, prior to Frege's work, the sentences of everyday proofs of arithmetic could be partially translated into Begriffsschrift in such a way as to make the correctness of the proof evident. Frege's *Begriffsschrift* contains a guide to the replacement of some of our everyday language with a more precise notation. The use of this notation to replace some of our everyday locutions is an important preliminary step in Frege's replacement of arithmetic. He says,

> We are very dependent on external aids in our thinking, and there is no doubt that the language of everyday life—so far, at least, as a certain area of discourse is concerned—had first to be replaced by a more sophisticated instrument, before certain distinctions could be noticed. (PW 195)

For instance, Frege's conditional-stroke and generality sign could replace some of the English words in "If a number is less than 2, its square is less than 4." The result, in modern notation, would be: $(x)(x < 2 \rightarrow x^2 < 4)$. This translation, along with some of Frege's logical laws, some general laws of arithmetic, and definitions of the less-than relation and the square function as well as definitions of '2' and '4' using '1' and the successor function, should allow us to exhibit the structure of the proof of the English claim from those definitions in such a way as to make its correctness apparent.

Such proofs, Frege suggests in section 6 of *Foundations*, are possible. He claims, in section 6, that every number can be defined from its predecessor. Consequently, any occurrence of a numeral in a sentence of everyday arithmetic can be replaced by a new sort of numeral, a name constructed in such a way as to indicate that the number named is the result of finitely many applications of the successor function to zero (e.g., $(((0 + 1) + 1) + 1)$). The result will be a stilted expression of whatever is expressed by the original expression. Furthermore, Frege argues in section 6 that, given the associativity of addition and the definitions '$0 + 1 = 1$', '$1 + 1 = 2$', and so on, the nonstilted expression should be provable from the stilted expression. In fact, the result of taking a proof in Peano arithmetic and substituting Frege's logical symbols for contemporary symbols (e.g., substituting Frege's conditional-stroke for the arrow) will be such a proof.[27] Such proofs are not, of course, Begriffsschrift proofs. For there will remain a number of non-logical symbols occurring in the proof—in particular, the numeral '1' and a symbol for the successor function—and the premises of the proof are neither laws of logic introduced in Frege's *Begriffsschrift* nor consequences explicitly derived from those laws.

Let us now assume that the sentences of everyday arithmetic could be replaced with expressions in which all obviously logical notions are expressed by the appropriate Begriffsschrift expressions and all other notions are expressed in everyday language. It remains to replace the everyday terms that are not obviously logical terms—the numerals, for instance.

How are the numerals to be replaced by logical terms? Since Frege's original demand was that the terms '1' and 'number' be defined, one might suspect that these terms will be defined from Begriffsschrift expressions and then will be used to provide definitions of the remaining terms of arithmetic. This is not quite right. In particular, when it comes to the actual definitions, Frege begins not with '1' but with '0'. In *Basic Laws*, Frege gives Begriffsschrift definitions of two terms, '∅' and 'ƒ', which are to play the role of names for zero and the successor function. These terms can then be used to replace all numerals in the stilted sentences described above.[28] Provided there are no more special

---

[27]This is not quite correct. For the associative law for addition is actually a derived result in Peano arithmetic. Thus, a proof in Peano arithmetic might contain a subproof of this law. Also, Frege's claim, in section 6 of *Foundations*, is, of course, false—not all number theoretic truths are provable in Peano arithmetic.

[28]It is probably important to point out that this is actually oversimplified and, strictly speaking, incorrect. Frege's Begriffsschrift notation is sufficiently complicated and foreign

arithmetical terms in these stilted sentences, the result is a Begriffs-schrift expression. By following this procedure, the everyday arithmetical claim '0 ≠ 1' can now be translated into Frege's Begriffsschrift. Furthermore, the translation appears to satisfy the constraints described above. Since the sentence '0 ≠ 1' is regarded, in pre-Fregean arithmetic, as expressing a truth of arithmetic, its translation must be provable and must have the applications of the original sentence. If we follow the translation strategy described above, we get the Begriffs-schrift expression:

$$\top \, \mathtt{Q} = \mathtt{X}$$

This is proved in section 103 of *Basic Laws* (GGA vol. 1 131–132). The final line of this proof not only is remarkably similar in appearance to the everyday sentence of arithmetic but has the same applications. If all occurrences of pre-Fregean sentences, in what we take to be the legitimate inferences of pre-Fregean arithmetic, are replaced by the sort of translation described above, the result will be expandable into a legitimate inference of Frege's systematic arithmetic.

The other symbols in the everyday example with which I began, the terms for the less-than relation and the square function, can also, on Frege's view, be defined appropriately by Begriffsschrift symbols. An appropriate definition is, of course, a definition that meets Frege's requirements and from which all appropriate expressions are provable. The expressions that must be provable are the Begriffsschrift translations of the sentences we regard as truths of pre-Fregean arithmetic. For example, since we regard '2 < 4' as expressing a truth of arithmetic, the corresponding Begriffsschrift expression must be provable. For any statement that is regarded as an everyday pre-Fregean truth about the

---

to the contemporary reader that I have decided not to give the actual details of Frege's definitions. Thus, for instance, the definition of 'x' looks nothing like 'f(Q)'. In fact, the symbol 'f' does not appear in the definition of 'x', although it is possible to prove that x immediately follows Q in the f-series. Frege also proves this in section 103 (GGA vol. 1, pp. 131–132). Furthermore, Frege also proves that f is single-valued (GGA vol. 1, 109–113, section 87). Thus, it follows that, using Frege's definite description operator, it is possible to define any number, say n + 1, from definitions of n and f. It is also important to note that Frege's use of the symbol 'Q' rather than simply '0' is *not* indicative of his believing that his definition of 'Q' does not exactly capture the content of our everyday symbol '0'. In fact, 'Q' is his symbol for the *Anzahl* 0—the number ascribed to a concept under which no object falls. Frege defines the counting numbers differently from the real numbers, and he uses the symbol '0' for the real number 0.

less-than relation, there should be an analogous provable Begriffsschrift expression. There is, in fact, an analogous relation in Frege's systematic arithmetic—the relation borne by one finite integer to another if the first precedes the second in the f-series beginning with ϕ. (The notions of finite integer and precedence in the f-series are, of course, defined in *Basic Laws*.) If '$<$' were introduced as an abbreviation for the Begriffsschrift expression picking out the relation described here and if our everyday numerals (2,3, etc.) were introduced as the obvious abbreviations, then the appropriate expressions would be provable. That is, we could prove '$2 < 4$', and so on. Frege's definitions in *Basic Laws* are also supposed to be rigged in such a way as to allow us to make the appropriate inferences for applications of arithmetic. For instance, for any x, we could prove the defined Begriffsschrift expression '$2 < x$' from the defined Begriffsschrift expressions '$2 < 4$' and '$4 < x$', and we could prove that if there are 2 Fs and 4 Gs, the number of Gs is greater than the number of Fs.

Although he did not do this for all function and relation symbols from everyday arithmetic in *Basic Laws*, Frege believed his proofs in *Basic Laws* showed this could be done. The purpose of *Basic Laws* was to prove a series of basic laws of arithmetic from which truths analogous to what look to be everyday truths of arithmetic could be proved.

Were Frege successfully to carry out the project I have described here, any everyday sentence that was regarded as expressing a basic truth of arithmetic could be, in effect, translated into a provable Begriffsschrift sentence. Everyday pre-Fregean arithmetical proofs would have analogue proofs in the new science. For the result of replacing each line in any pre-Fregean proof which makes use only of basic truths of arithmetic and general logical laws with its analogous Begriffsschrift expression would be to transform the pre-Fregean proof into a Begriffsschrift proof by filling in all the gaps that were permitted by the informality of pre-Fregean arithmetic. Furthermore, the initial constraints placed by pre-Fregean arithmetic on Frege's definitions would guarantee that the new arithmetic has all the applications of pre-Fregean arithmetic. Thus, the only effect the replacement of arithmetic by Frege's new science would have on mathematical practice would be to make its proofs more cumbersome and more precise. Consequently, as the passage from "Logic in Mathematics" suggests, there seems to be no harm in replacing the terms Frege introduces with numerals and function-symbols from everyday arithmetic. The result would be expressions in the new science which look like expressions from the old science. The only thing, Frege says, that we are not permitted to do is to assume that the senses of these expressions are identical to the senses the expressions expressed in pre-Fregean arithmetic. Finally, the truths proved in this new science of arithmetic would be analytic—the ultimate grounds of their justification would be the logical source of knowledge.

This account of Frege's project suggests two obvious questions. First, is Frege trying to reform the practice of mathematics? In particular, does his introduction of a precise new science of arithmetic with proper epistemological foundations obligate us to leave off doing everyday arithmetic and to formulate all subsequent arithmetical proofs and applications in Begriffsschrift notation? Second, if the senses of the terms of Frege's new science of arithmetic are not the pre-Fregean senses these terms have carried, in what way has Frege shown that *our* arithmetic is analytic?

There is one reform that Frege does require of mathematical practice. This reform has to do with the requirements on definitions. I will return to this later. However, it is not clear that Frege's rigorization of arithmetic is meant to convince mathematicians to provide rigorous Begriffsschrift proofs in their research. For the actual practice of mathematicians is constrained also by psychological considerations. Although Frege maintains throughout his career that ideas and psychological laws are no part of logic, this does not prohibit him from acknowledging that we might be incapable of judging except in accord with psychological laws. He says, for instance,

> The task in hand is precisely that of isolating what is logic. This does not mean that we want to banish any trace of what is psychological from thinking as it naturally takes place, which would be impossible; we only want to become aware of the logical justification for what we think. (PW 5)

Actual Begriffsschrift proofs are sufficiently cumbersome that it may be impossible to use them in mathematical research. Even in *Basic Laws* Frege introduces abbreviations and shortcuts. What, then, are we to make of Frege's elaborate definitions of the terms used in arithmetic? Frege says, in "Logic in Mathematics,"

> We simply do not have the mental capacity to hold before our minds a very complex logical structure so that it is equally clear to us in every detail. . . . And yet we can still draw correct inferences, even though in doing so there is always a part of the sense in penumbra. (PW 222)

He goes on to say that Weierstrass, who, in Frege's estimation, had false views about what numbers are, nonetheless made correct inferences. How can this be?

It is important to note that Frege has no objection to our using, without re-proof, something that has already been proved. Thus, if we could, for instance, prove the axioms of Peano arithmetic by means of Frege's work in *Basic Laws*, we could then derive further truths by means of these axioms and general logical laws. The status of these further proofs is not dependent on the knowledge or understanding of the person who constructed the proof. It is necessary only that the Begriffsschrift proofs exist, for otherwise the further proofs would not have adequate justification. Of course, one might want to say that the truths proved in Begriffsschrift are not really those expressed by the axioms of Peano arithmetic. Frege says, of Weierstrass, "His sentences express true thoughts, if they are rightly understood" (PW 222). Thus, given the accord between Frege's new science of arithmetic and pre-Fregean arithmetic—that is, given that more or less the same sentences appear as the final lines of proofs—Frege need not require us to provide actual Begriffsschrift proofs of claims about number. The new science can be carried out more or less as the old science was carried out. What is important is that it be agreed antecedently that the terms of arithmetic are to have the Bedeutung assigned them by Frege's definitions. Should any confusion or dispute arise, it may be of use to bring out these definitions. We must regard everyday sentences and proofs of arithmetic as abbreviations of precise statements and proofs if we are to view arithmetic as having a proper epistemological foundation.

What of the second question? I have said that the same sentences appear as final lines in proofs of both the pre- and post-Fregean sciences. But Frege also says, "A mere wording without a thought content can never be proved" (CP 316). One might suppose, given that his new everyday sentences cannot be regarded as having the same sense as the old everyday sentences they replace, that his epistemological foundation is not a foundation for pre-Fregean arithmetic.

It is not immediately obvious how this issue can be addressed. For, given Frege's strict requirements for admissibility of terms for science, it is not clear that there is any respect, except for their being regarded as abbreviations of Begriffsschrift expressions, in which pre-Fregean sentences can be viewed as expressing something provable or having an epistemological foundation. This difficulty does not arise, however, if we talk about the senses of expressions of pre- and post-Fregean arithmetic. For a term need not be admissible for science in order to have a sense—fictional names, for instance, have sense but no Bedeutung. Frege suggests that every grammatically well-formed proper name has a sense (CP 159). Furthermore, there seems to be an important relation between the senses of expressions of pre- and post-Fregean arithmetic. Although the senses of pre-Fregean expressions are insufficiently clear and unambiguous for Frege's scientific purposes, there are some respects in which these senses are clear. In those respects, the clear, unambiguous senses Fregean arithmetic assigns to these expressions agree with the pre-Fregean senses. That is, Fregean arithmetic contains all the scientific content of pre-Fregean arithmetic. To see this, let us consider some of Frege's remarks about the notion of sense, particularly the sense of an assertoric sentence.

In his introduction of sense in "On Sense and Meaning," he characterizes sense as the mode of presentation of the thing designated by an expression. In the case of an assertoric sentence, the thing designated is its truth-value. He also characterizes sense as what is expressed by an expression—in the case of a sentence, the thought it expresses or its cognitive value. But Frege devotes very little discussion to how this is to be understood.[29] His central worry about how these characterizations might be misunderstood is that his readers might infer that the sense of an expression is constituted, in whole or in part, by the ideas individuals associate with it. Most of his paper is devoted, not to fleshing out the notions of sense and Bedeutung, but to an attempt to convince the reader that it is not unreasonable to say that the truth-value of a sentence is its Bedeutung.

There are two general sorts of characterization of Frege's notion of

[29]Although Frege introduces his notion of sense by talking about the senses of complex proper names, his remarks about the sense of proper names are brief and unhelpful. The extent to which Frege's remarks are confusing becomes most apparent when everyday proper names are under consideration. The most explicit remark seems to be the footnote about the sense of 'Aristotle' near the beginning of "On Sense and Meaning" (CP 158). This footnote has been taken to suggest that the sense of a proper name is given by a definite description that refers to the same object (Peter Geach, "Frege," in Three Philosophers; Saul Kripke "Naming and Necessity," 255; Currie, Frege: An Introduction to His Philosophy, 170). Dummett points out that Frege never explicitly said this and says it is dubious that Frege supposed such a thing (FPL 110). David Bell suggests, in Frege's Theory of Judgement (64), that the sense of a proper name is not something that can be stipulated but is something that can only be hinted at, and that a proper name's sense "is that it purports to refer to a determinate object of a given sort with which it has been conventionally correlated."

sense. The one mentioned above, the mode of presentation or cognitive value, looks to be derived from an understanding of a central feature of everyday language. It is easy to regard sense, on this characterization, as what is understood when an expression is understood. But it is important to note that this is correct only insofar as what is understood is relevant to determining truth-values of expressions of which the expression is a part. Thus, Frege suggests that 'although', 'but', and 'yet' have the same sense as 'and', although subsidiary clauses beginning with one of the first three are thereby illuminated in a peculiar fashion (CP 172). While this illumination is surely understood by speakers of the language who understand 'although', 'but', and 'yet', it is not part of the senses of these words. Given Frege's understanding of logic, what is relevant to determining truth-values is also what is relevant to evaluating inferences. It is, for instance, the sense of 'and' which licenses the inference from '*A* and *B*' to '*A*'.

He characterizes sense, second, as part of what, in *Begriffsschrift*, is called 'judgeable content' (*beurtheilbarer Inhalt*). This *Begriffsschrift* notion, he says, in a letter to Husserl (COR 63), in "On Concept and Object"—a paper written as a companion to "On Sense and Meaning" (CP 187)—and in *Basic Laws* (BLA 38), has been divided into thought (the sense of an assertoric sentence) and truth-value (its Bedeutung). In *Begriffsschrift*, Frege says, of judgeable contents, "In a judgement I consider only that which influences its *possible consequences*. Everything necessary for a correct inference is expressed in full, but what is not necessary is generally not indicated" (BEG 12). Since the sense of the parts of an assertoric sentence will be parts of the thought expressed, the sense of an expression is bound up in what is relevant for evaluating inferences or determining truth-values.[30]

How are the senses of pre- and post-Fregean arithmetic related? Although the laws of logic do not properly apply to pre-Fregean arithmetic, the view of pre-Fregean arithmetic as a science is based on the assumption that certain inferences of pre-Fregean arithmetic, properly understood, are correct. Furthermore, the constraint on the replacement of pre-Fregean sentences with Fregean sentences is that the pre-

[30]There is a serious problem that I have not mentioned and that, for the time being, I will not address. It seems, from what I have said here, that sameness of sense for assertoric sentences ought to amount to logical equivalence. In fact, Frege says something like this in a letter to Husserl written some time after "On Sense and Meaning" (COR 70). But this seems to conflict with the notion of sense as cognitive content. I will discuss this in chapter 6.

Fregean inferences that were accepted by mathematicians must be con-structible in Fregean arithmetic. If what is important to the sense of a sentence is the correct inferences in which it can appear, the senses of Fregean sentences of arithmetic capture what is clear about the senses of pre-Fregean sentences of arithmetic. Thus, it does not seem unrea-sonable to say that the sense of pre-Fregean sentences of arithmetic is included in the Fregean understanding of these sentences. Conse-quently, there is a significant respect in which the purely logical proofs of these sentences can be viewed as justifying pre-Fregean arithmetic.

Viewed in this way, there is no reason to suppose that Frege's Be-griffsschrift definitions of the terms of arithmetic are the *only* definitions that will allow gapless proofs of the truths of arithmetic. What is impor-tant is simply to show that there is some way of replacing pre-Fregean arithmetic with a systematic science. This is entirely in line with his comments about the justification of calculation with complex numbers. He says:

> Well, perhaps it is indeed possible to assign a whole variety of different meanings to $a + bi$, and to sum and product, all of them such that those laws continue to hold good; but it is not immaterial whether we can or cannot find *some* such a meaning for those expressions.[31] (FA 111)

And, more metaphorically:

> In a geometrical theorem where a constructed line is used for the proof, the auxiliary line does not occur in the theorem. Perhaps more than one such line is possible, as for instance, where we can select a point at will. But however much we can dispense with each and any of them individu-ally, still the cogency of our proof depends on its being possible to draw some line of the required character. (FA 112)

Furthermore, even if there were other definitions making use of geo-metrical notions, which would allow us to replace pre-Fregean arithme-tic with a new science, this would not show that pre-Fregean arithmetic

---

[31]This differs from the Austin translation. The occurrence of 'meaning' in the last sentence of this passage replaces Austin's 'sense'. This replacement is not due to a disagreement with Austin's translation, since the German word is *Sinn*. Rather, this is a response to a remark in a letter from Frege to Husserl. Frege says, in an 1891 letter to Husserl, that, having drawn the sense/Bedeutung distinction, he would now replace the many occurrences of 'sense' in sections 100, 101, and 102 with 'meaning' (*Bedeutung*). See COR 63/WB 96. The passage quoted here is from section 101, and I believe that the substitution of 'meaning' for 'sense' here is one of those Frege had in mind.

was not analytic. If Frege's *Basic Laws* is successful, he has shown via his proofs that no use of the intuition is required for the justification of the proofs of pre-Fregean arithmetic. The upshot is that Frege's definitions of the numbers does not amount to a precise statement of the Bedeutung of pre-Fregean arithmetic.

It should be noted that, although the project I have just described is not generally attributed to Frege, it is not entirely unfamiliar. In fact, such a project is described by Carnap (one of Frege's students) in the introduction to *Logical Foundations of Probability*. He says there:

> The task of explication consists in transforming a given more or less inexact concept into an exact one or, rather, in replacing the first by the second. Perhaps the form 'explicans' might be considered instead of 'explicatum'; however, I think the analogy with the terms 'definiendum' and 'definiens' would not be useful because, if the explication consists in giving an explicit definition, *both the definiens and the definiendum in this definition express the explicatum, while the explicandum does not occur.*[32]

Frege's definitions of 'one' and 'number' are explications in this sense. 'One' and the complex expression that Frege uses to define it have the same sense, but they have the same sense only by virtue of the definition. The definition does not consist of an analysis of what was previously understood by 'one', and it should be viewed as nothing more nor less than an abbreviation. While the purpose of Frege's definitions is to provide foundations for arithmetic in the sense described above—to allow us to construct a proof, for each sentence that had previously been taken to express a truth of arithmetic, whose last line is that sentence—the fact that Frege's definitions are formulated for this purpose is only a part of a hidden agenda; it is in no way a part of the actual definition. Frege says,

> In constructing the new system we can take no account, logically speaking, of anything in mathematics that existed prior to the new system.

---

[32]Rudolf Carnap, *The Logical Foundations of Probability*. This similarity has also been noted by Michael Resnik in *Frege and the Philosophy of Mathematics*, 184. Resnik's point, however, is rather different from mine. Resnik uses this comparison to suggest that Frege's definitions do not preserve pre-Fregean sense and that there is some reason to believe that they also do not preserve pre-Fregean reference. I have argued, there is an important respect in which Frege's goal can be characterized as the preservation of what pre-Fregean sense was associated with the terms he is defining. Furthermore, as I read Frege, it makes no sense at all to talk of the pre-Fregean reference of these terms, since there was none.

> Everything has to be made anew from the ground up. Even anything that
> we may have accomplished by our analytical activities is to be regarded
> only a preparatory work which does not itself make any appearance in the
> new system itself. (PW 211)

From this point of view, Frege goes on to say, definitions are to be
regarded as nothing more than arbitrary stipulations.

What, then, are we to make of Frege's demand that definitions be
justified? There are two sorts of justification which appear to be necessary for definitions. A discussion of the first appears as a criticism of
mathematical practice. The mathematical practice in question is that of
giving definitions that are "justified only as an afterthought, by our
failing to come across a contradiction" (FA ix). Later, Frege criticizes a
proposed definition of '$a + b$' because

> $a + b$ would be an empty symbol if there were either no member or several
> members of the basic series which satisfied the prescribed condition. That
> this does not in fact ever happen, GRASSMANN simply assumes without
> proof, so that the rigour of his procedure is only apparent. (FA 8)

The sort of justification Frege insists on here, however, is required only
of the mathematicians whose work is presystematic. For, when Frege
formulates the ultimate version of his Begriffsschrift, in *Basic Laws*, his
principles for forming legitimate names are designed to preclude the
possibility of a contradiction and to guarantee that proper names pick
out exactly one object, and that concept names determinately hold, or
do not, of each object. As he says, in "On Sense and Meaning," "A
logically perfect language [Begriffsschrift] should satisfy the conditions
that every expression grammatically well-constructed as a proper name
out of signs already introduced shall in fact designate an object" (CP
169). Thus, if a system is constructed using a logically perfect language,
there is no need to justify definitions. And Frege's definitions are offered as part of his systematic science whose language (Begriffsschrift)
is logically perfect.

The other sort of justification Frege seems to demand (and provide) is
also presystematic. This concerns his particular choice of definitions.
Why does Frege's definition of the number 'one' have more legitimacy
than a definition on which 'one' stands for, say, the moon? The point of
Frege's definition is to allow him to formulate a system that can replace
arithmetic. Such a replacement will only be possible if the sentences of
the systematic science corresponding in shape to sentences in the pre-

systematic science which were (presystematically) taken to be truths of arithmetic are themselves true. Frege's presystematic justifications of his definitions (in particular, the sections of *Foundations* labelled "Our definition completed and its worth proved") are designed to convince his readers that his systematic arithmetic can replace ours. This justification, again, will play no role in his systematic science.

From the point of view of Frege's systematic science of arithmetic, his definitions of '1' and of the concept of number are nothing more nor less than mere abbreviations. But from the point of view of Frege's epistemological views, the existence of these definitions is significant. The significance of the particular definitions offered in Frege's *Basic Laws* is that they are meant to allow him to build a purely logical science that can be used for all the purposes for which presystematic arithmetic is used. Frege believes that such a construction will show that the truths of arithmetic are analytic. But, as I have argued above, Frege's construction of a systematic science of arithmetic from general logical laws can only show that the truths of (our) arithmetic are analytic if we regard not only our arithmetic but virtually all scientific research—except for proofs in Euclidean geometry—as seriously defective. Euclidean geometry, on Frege's view, is not defective because its primitive terms are actually epistemologically primitive (they are primitive terms of our intuition of space and, accordingly, symbols for these terms can be directly introduced into Begriffsschrift) and the axioms of geometry are primitive truths of inner intuition. The only real defects in the proofs of Euclidean geometry is that they are not rigorously expressed in Begriffsschrift notation.

I have argued that the tension between Frege's view that definitions of the number 'one' and concept of number are required for the legitimacy of our arithmetic and his view of definitions as purely stipulative creates a need for a distinction of the sort drawn in "On Sense and Meaning." This distinction allows Frege to explain both what is wrong with our everyday understanding of the basic terms of arithmetic and what features of our everyday understanding of arithmetic are of value and must be preserved in a systematic science that can replace arithmetic. The defect of our everyday understanding of these terms is that it is insufficiently precise. These terms, as they are understood, lack Bedeutung. The characterization of Frege's project of defining these terms which I have provided in this chapter highlights some of the difficulties inherent in translating 'Bedeutung' either by 'meaning'—as it is now customarily translated—or by 'reference'.

One of the features of Frege's distinction which allows him to say what is wrong with our everyday understanding of arithmetic is that the requirements for a term's having Bedeutung are not met by most of our everyday linguistic expressions. A concept expression, for instance, has Bedeutung just in case it determinately holds or fails to hold of each object. But this is not true of our everyday concept terms. In general, they only apply over limited ranges—only particular sorts of objects, for instance, can be said to be bald or not to be bald—and have vague boundaries. Even such apparently precise mathematical concepts as that which holds of two objects if the first is greater than the successor of the second turn out not to satisfy Frege's sharp boundary criterion for admissibility in science or having Bedeutung. For who would claim that there is an answer to whether or not the moon is greater than the successor of the sun? The translation of 'Bedeutung' by 'meaning' will require us to attribute to Frege the view that our everyday concept-expressions are meaningless. In addition, given Frege's standards, our everyday definite descriptions, as well as the numerals, cannot have had Bedeutung before his work. The translation of 'Bedeutung' by 'reference' will require us to attribute to Frege the view that not only the pre-Fregean numerals but also most of our everyday proper names are referenceless.

I have chosen to leave 'Bedeutung' untranslated in my discussions of these consequences in order to avoid attributing obvious absurdities to Frege. Given that one of Frege's aims in "On Sense and Meaning" is to introduce a term ('Bedeutung') to describe a feature that is required of every term of a systematic science, there is no reason to saddle Frege with these absurdities. The absurdities that result from construing Frege's use of this term via our everyday understanding of 'meaning' and 'reference' do not follow if Frege is introducing a technical use of a familiar term. And the absurdities that result from construing Frege's use of 'Bedeutung' as identical to our technical, late twentieth-century understanding of these terms provide an argument that—in spite of the fact that our own use has its historical roots in Frege's texts—it is a mistake to interpret Frege's use of 'Bedeutung' in this way. There is no harm in using either 'meaning' or 'reference' to translate 'Bedeutung' as long as the word used is clearly marked as a technical Fregean term rather than an everyday term, and as long as we do not assume that Frege's technical use is identical to our contemporary technical use.

For Frege, showing that a term has Bedeutung, on the account presented here, seems to amount to showing we have the appropriate sort of definition of this term from primitive terms. One purpose of showing that a term has Bedeutung is to exhibit the epistemological foundation of the science in which it is used. The other is to guarantee that there is no ambiguity or vagueness in our understanding of the term. Frege seems to be assuming that no work is required to show that primitive terms have Bedeutung. This is understandable if establishing that a term has Bedeutung is nothing more than exhibiting epistemological roots and eliminating vagueness.

To see this, consider Frege's discussion, in chapter 1 of *Foundations*, of Kant's and Mill's views about the numbers. Frege argues that Kant's and Mill's attempts to characterize the truths of arithmetic as, respectively, synthetic a priori and synthetic a posteriori fail. One explicit consequence of the argument is that there is not universal agreement about what the numbers are. Another explicit consequence is that it is not clear that numbers are properties or objects belonging to the physical world or belonging to our inner intuition. Frege adds an argument, in sections 15–17, that it is not immediately evident that arithmetic is analytic. I have argued that the exhibition of these explicit consequences is designed to convince his audience that the numeral '1' and the concept-word 'number' cannot be primitive terms. I have also argued that, on Frege's view, to show that a term is not primitive, it suffices to show that there is not universal agreement about its Bedeutung or that it is not immediately obvious what the ultimate grounds of our justification of the thoughts expressed by sentences using this term are. Hence, primitive terms cannot be vague and their epistemological status (that is, the sort of ultimate grounds which can license truths that are expressed by sentences in which these terms appear) cannot be obscure.

None of this information amounts to an account of Frege's notion of primitiveness. But does Frege need such an account? His aim is to show that the everyday science of arithmetic can be replaced by a systematic science of pure logic. In order to do this, it is necessary to define the terms of arithmetic from terms of logic and then to show that the truths of arithmetic can be proved from these definitions and general logical laws. As long as the logical terms used in Frege's definitions are recognizably and uncontroversially terms of logic, and as long as he can show that the appropriate sort of proofs can be given, no account of primitiveness is necessary for the justification of his claim to have shown that the truths of arithmetic are analytic. It might seem that an account of primitiveness would be most helpful in the arguments that precede and are meant to motivate the construction of Frege's system—the arguments that, for instance, it would be wrong to take '1' as a primitive term. But these arguments depend less on our understanding Frege's notion of primitiveness than on our understanding his epistemological presuppositions. Frege simply does not need an account of primitiveness.

I have argued that, given Frege's overall project, it is entirely appropriate for him to introduce a term for a kind of content whose significance is exhausted by the demarcation of sharp boundaries of an expression and the indication of the epistemological roots of truths expressed by sentences in which the expression appears. To show that an expression has this sort of content, it suffices to show that it is either primitive or definable in the appropriate way from primitive terms. Frege's word for this is 'Bedeutung' and hence his use of this word should not be conflated with our contemporary use of 'reference'. But there is something odd about this picture of Frege's project.

It is no accident that our contemporary use of 'reference' has its historical roots in Frege's use of 'Bedeutung'. The purpose of Frege's epistemological work seems to be the demonstration that we really can have objective knowledge of the truths of arithmetic. On the other hand, if Frege's notion of objective knowledge is related to the notion of Bedeutung described above, it may seem that objective knowledge need have no relation to a mind-independent reality. For the requirement that all terms of a sentence have Bedeutung looks to be nothing more than a requirement that our linguistic terms be precise. There is no obvious reason to think that a precise expression must describe an external reality. What looks to be required, given Frege's ambitions, is something like a reference relation in a contemporary sense—a hook between words and the world. I will address this issue in chapter 4.

PART II

CHAPTER 4

Bedeutung and Objectivity

AT this point it will help to reconsider some of the difficulties with Frege's use of the term *Bedeutung*. I have argued that, when this term is translated as 'meaning', many of the consequences of Frege's views seem absurd. Surely one would not want to claim, for instance, that pre-Fregean arithmetic was meaningless. This is not merely a problem with translation, however. The translation only serves to make the problem more vivid.

Before his formulation of the sense/meaning distinction, Frege viewed sentences only as having (or not having) judgeable contents. Defective sentences, assertoric sentences that were not assigned sufficiently precise content, could not have judgeable contents. There is no possibility, in Frege's pre-1891 work, of a sentence without judgeable content playing any role at all in communication and science. It looks as if these sentences must count as gibberish, as if Frege truly is committed to viewing them as meaningless. But Frege is committed to the view that we *do* communicate prescientifically. What is the effect of Frege's introduction of the sense/Bedeutung distinction on this uncomfortable situation? This distinction divides judgeable content into thought and truth-value. Defective sentences, do not, of course, have truth-values. The judgement that a sentence expresses a true thought should be made only when all its terms are sufficiently precise. But it is not clear that a defective sentence (a sentence that is inadmissable for science) cannot express a thought.

First of all, there are clearly sentences that have sense (express a thought) but no truth-value. Frege gives, as an example, the sentence "Odysseus was set ashore at Ithaca while sound asleep" (CP 162). Since 'Odysseus' has no Bedeutung, the sentence is neither true nor false. Frege says, "The thought remains the same whether 'Odysseus' means something or not" (CP 163). Frege repeatedly gives expressions used in fiction as examples of expressions with sense but no Bedeutung. Thus, an expression can have content without Bedeutung. Furthermore, it seems that this situation can arise in presystematic scientific contexts as well as in fictional contexts. Frege says,

> It may perhaps be granted that every grammatically well-formed expression figuring as a proper name always has a sense. But this is not to say that to the sense there also corresponds a thing meant. The words 'the celestial body most distant from the Earth' have a sense, but it is very doubtful if there is also a thing they mean. (CP 159)

If the expression Frege uses here has a sense, then so must the sentence "The celestial body most distant from the Earth is a celestial body". If Frege's doubts are right and the expression has no meaning (Bedeutung), then the sentence in which it appears, like the fictional sentence, has no truth-value but expresses a thought.

If these comments are applicable to presystematic science, Frege may be able to say that sentences of presystematic science can have some sort of content. However, it seems that Frege explicitly denies this. He says, in a later essay titled "Compound Thoughts," "I call any sentence a sentence proper if it expresses a thought. But a thought is something which must be either true or false, *tertium non datur*" (CP 392). What are we to make of the apparent contradiction between this claim and his claim from "On Sense and Meaning" that fictional sentences express thoughts? It is, of course, conceivable the Frege changed his mind. But I think it is more likely that, in the passage from "Compound Thoughts," Frege is meaning to be talking about scientific contexts. This is supported by a number of other passages in which the same phrases occur. For instance, in "Foundations of Geometry II," Frege says, "Now a real proposition expresses a thought. The latter is either true or false: *tertium non datur*" (CP 329). He adds the following in a footnote to this sentence: "For we are here in the realm of science. In myth and fiction, of course, there may occur thoughts that are neither true nor false but just that: fiction." These claims also appear in some unpublished notes about logic (PW 186). But it is not clear how helpful this reinterpretation of the passage from "Compound Thoughts" is. Although Frege no longer is obviously contradicting the passage from "On Sense and Meaning," presystematic science is hardly fictional. How can these passages be reconciled with the claim that sentences of presystematic science express thoughts?

It is difficult to see a principled way of arguing that fictional sentences express thoughts but sentences of presystematic science do not. For the difference between fiction and presystematic science has to do not with the nature of what is expressed by their sentences, but with the standards to which we hold what is expressed. Many of the same expressions are used in fictional and presystematic scientific contexts. The difference is that we have more rigorous standards for what is acceptable as a scientific assertion. Furthermore, some of Frege's examples, in "On Sense and Meaning," of expressions that have sense but no meaning are expressions from presystematic science. The solution, I think, is to read the above passages as stating, not what is characteristic of all thoughts but, rather, what is required of thoughts that are acceptable for purposes of logic or systematic science. In "Logic in Mathematics," he says, "In myth and fiction thoughts occur that are neither true nor false. Logic has nothing to do with these. In logic it holds good that every thought is either true or false, *tertium non datur*" (PW 198). The terms of a presystematic science are insufficiently precise and hence not acceptable for purposes of logic. Thus, it may be that, in the contrast between science and fiction, presystematic science belongs to the realm of fiction. This is also suggested by a later passage from "Logic in Mathematics," where Frege says, "Now the idealist may say that it is wrong to hold that the name 'Etna' designates something. If that were so, the speaker, whilst believing himself to be operating in the realm of truth, would be lost in the realm of myth and fiction" (PW 232). If someone could take herself or himself to be engaged in scientific inquiry (that is, to be concerned with truth) but, in fact, be "lost in the realm of myth and fiction," then there is no clear distinction to be made between what is expressed by sentences of myth or fiction and what is expressed by presystematic science. It is not that sentences of presystematic science do not express thoughts but, rather, that the thoughts expressed are defective—they are insufficiently precise to determine truth-values. Frege goes on to say,

> Now if we were concerned only with the sense of 'Etna is higher than Vesuvius', we should have no reason for requiring that the name 'Etna' should have a meaning as well; for in order that the sentence have a

sense, it is only necessary for the name 'Etna' to have a sense; the meaning contributes nothing to the thought expressed. If therefore we are concerned that the name 'Etna' should designate something, we shall also be concerned with the meaning of the sentence as a whole. That the name should designate something matters to us if and only if we are concerned with truth in the scientific sense. So our sentence will have a meaning when and only when the thought expressed in it is true or false. (PW 232)

In this passage, Frege seems to indicate that it is possible to be concerned with the sense of a sentence that belongs to a presystematic science and that such a sentence might express a thought that is neither true nor false. Frege says something similar in a letter to Russell when, in a discussion of the thought expressed by '$7 - 1 = 6$', he says, "Thus the sense is independent of whether there is a meaning" (COR 165). Finally, in "Thoughts," Frege twice says that if he is using an expression that he takes to have a meaning but that, unbeknownst to him, does not, then he has inadvertently lapsed into fiction (CP 362, 367). Researchers in a presystematic science are in precisely the situation Frege describes here. For, although they assume that the terms they use have meaning, these terms fail to have meaning due to their imprecision and/or ambiguity. Thus, Frege's division of judgeable content into thought and truth-value makes it possible to view the sentences of presystematic science as expressing thoughts and yet as being defective from the point of view of systematic science.

There still seems to be a problem with this account of the defect.[1] It is a consequence of Frege's views that most presystematic sentences had no truth value. But 'true' is not simply a term of art for Frege. Indeed, Frege says this term is simple and indefinable.[2] Frege's laws of logic are laws of truth. An understanding of this term underlies the activity of judging. And truth is connected with evaluation. Even after Frege's introduction of a systematic science to replace everyday arithmetic, we would still want to be able to say that, for instance, someone who had earlier claimed that $132 + 97 = 229$ was right while someone who had earlier claimed that $132 + 97 = 199$ was wrong.

In order to understand how Frege could respond to this, it is impor-

---

[1] I am indebted to an anonymous reviewer for suggesting that this might be a problem.
[2] See the beginning pages of "Thoughts" (CP 351–352), where this view is most fully articulated. The claim that the notion of truth is primitive is also made in some of Frege's unpublished notes about logic (see, for instance, PW 126, 129), and the introduction of his notation in *Basic Laws* makes use of the notion of truth without definition.

tant to remember that Frege's project is to show how it is that we can have knowledge of arithmetic. Our everyday assumption is that we do have knowledge of truths of arithmetic and that it is true (for instance) that 132 + 97 = 229. But Frege wants to argue that this is a *mere* assumption. Because of his epistemological views, he believes that we need more precision both in our understanding of the terms used—he thinks that we are not really clear about what we are talking about—and in our justification. His discussion of Kant's and Mill's views, in chapter 1 of *Foundations*, is designed to show us that our use of numerals really is not clear. For, were it clear, the epistemological source of our knowledge would be evident. He is trying to address the problem that "the first prerequisite for learning anything is thus utterly lacking—I mean, the knowledge that we do not know" (FA iii). Frege's philosophical picture provides a story about what work is necessary in order to ensure that our use of arithmetical terms is sufficiently clear and that the truths can be justified from ultimate grounds. In order to show that we can have knowledge of arithmetic, it is necessary to replace our arithmetic with a systematic science. In the context of pre-Fregean arithmetic, we claim that '132 + 97 = 229' and '132 + 97 ≠ 199' are true. A systematic science that can replace arithmetic will allow us to construct proofs with these sentences as their final lines. The vindication of the pre-Fregean claims that 132 + 97 = 229 and 132 + 97 ≠ 199 is the existence of a systematic science of arithmetic in which '132 + 97 = 229' expresses a truth and '132 + 97 = 199' does not.

This vindication is not trivial. If our arithmetic could not be replaced by a systematic science, that is, if those sentences which we have regarded as expressing truths of arithmetic cannot be replaced by sentences expressing truths in Frege's sense (truths that can be proved from primitive truths), this would show that the defects of our arithmetic go beyond the vagueness of its terms. Our arithmetic, in this situation, would be wrong.

Of course, there is an obvious problem with this. Frege's proofs in *Basic Laws* were dependent on an axiom that made his logic inconsistent. But, although Frege's attempt to replace arithmetic with a systematic science failed, he did not infer that pre-Fregean arithmetic was wrong. This does not, by itself, undermine the above description of Frege's project. For the failure of one attempt to systematize arithmetic does not show that arithmetic cannot be systematized. In fact, Frege did not give up his assumption that it was possible to replace arithmetic with a systematic science. Rather, since he became convinced that arithmetic could not be replaced by a systematic science whose ultimate ground is the logical source of knowledge, he inferred that it must be possible to replace arithmetic with a science whose ultimate ground is inner intuition.[3]

Finally, it is important to note that Frege's view does not preclude the existence of an alternative arithmetic that can also be replaced by a systematic science and on which $132 + 97 = 199$. It is by no means clear that there is such an alternative arithmetic that amounts to anything more than a notational variant of our own. And there was no pre-Fregean alternative arithmetic. However, were there an alternative pre-Fregean arithmetic that could be replaced by a systematic science, in Frege's sense, and on which $132 + 97 = 199$, it is unlikely that Frege would want to say that users of the alternative arithmetic who claim that $132 + 97 = 199$ were wrong. For those people, Frege's Begriffsschrift definitions of the numerals, as well as signs for addition and equality, would not work. Were their arithmetic replaced by a systematic science, they would thereby have associated with their symbols—the numerals, '+', and '='—different senses and meanings from the senses and meanings we associate with the same symbols. These people would not be wrong; they would, in effect, be using a distinct language.

Something still looks to be missing on this account: an explanation of how science can be about the world. The task of identifying primitive terms for a science, developing definitions from primitive terms for all other terms needed, and showing how truths about notions for which the complex terms are used can be derived from expressions of truths in which only primitive terms appear seems suspiciously internal. How can achieving the precision Frege demands suffice to establish epistemological foundations for a science? If the defects of presystematic science can be characterized as their having fictional thoughts, what makes systematic science nonfictional? Why, for instance, should a science of elves and fairies be precluded?[4]

The short answer is that such a science is not precluded. From a precise definition of what it is to be an elf or a fairy it would, at the very

[3]See the 1924–25 notes titled "Sources of Knowledge of Mathematics and the Mathematical Natural Sciences" and "Numbers and Arithmetic" (PW 267–277).

[4]The science of elves and fairies might seem farfetched. Perhaps it would be more reasonable to consider the possibility of Frege's being committed to the view that the primitive terms of some well-worked-out medieval theology must have Bedeutung. But I do not see how the source of our knowledge of such primitive terms could be a source Frege recognized, that is, logical, inner intuition, or the senses.

least, be possible to make inferences about the nature of elves or fairies. All of these inferences would be purely logical inferences from definitions, hence their consequences would be analytic truths. But if I seem to have suggested that the existence of an a priori science of elves and fairies follows from Frege's views, it is important to be clear about the nature of this science. Frege says repeatedly that there is nothing wrong with the scientific employment of concepts under which no object falls.[5] It does not follow from the permissibility of scientific use of the concepts of elfhood and fairyhood that there *are* elves and fairies, any more than it follows, from the permissibility of the scientific use of the concept *fraction smaller than 1 and such that no fraction smaller than 1 exceeds it in magnitude*, that there is such a fraction. In fact, Frege says that this concept is "quite unexceptionable" but that no object falls under it (FA 87–88). To show that there *are* elves or fairies, it would be necessary to come up with an actual object that satisfied the precise criteria for fairyhood or elfhood. Unless this could be achieved, the analytic science of elves and fairies described above would be nothing more than a word game.

[5]See FA 87–88, 105; BLA 11–12; CP 133, 134, 205, 225, 227; PW 34, 122, 179. This may seem somewhat puzzling in light of this remark of his in *Foundations*: "It is only in virtue of the possibility of something not being wise that it makes sense to say 'Solon is wise'. The content of a concept diminishes as its extension increases; if its extension becomes all-embracing, its content must vanish altogether" (FA 40). This is especially puzzling because Frege sometimes uses the term *Inhalt* ('content') interchangeably with *Sinn* or *Bedeutung* in *Foundations*. But it is clear that Frege's use of 'content' here is different, for he talks about the content of a concept rather than the content of a concept-expression. Frege's point here is not that a concept that is true either of everything or of nothing fails to be a concept. It is, rather, that once we realize either that a concept has everything or has nothing in its extension, it is uninformative to say that a particular object falls (does not fall) under the concept. In particular, to take 'one' as a concept-expression that holds of all objects would be, in effect, to make it impossible to use the concept-expression 'one' in an informative way. This would make it impossible to explain the content, for instance, of the claim that Frege had one horse. As Frege goes on to say, "It is not easy to imagine how language could come to invent a word which could not be of the slightest use for adding to the description of any object whatsoever" (FA 40). As a matter of fact, it is not at all clear that there are words that are meant simply as contradictory concepts. Contradictory concepts, generally, are formed from other concept-expressions that are not contradictory or tautological. And Frege says, in *Foundations*, that such contradictory concepts as 'square circle' and 'wooden iron' are "not so black as they are painted" (FA 87). This is not an idle remark or a mistake. For Frege makes use of an obviously contradictory concept (the concept of not being self-identical) in his definitions of the numbers and concept of number. Thus, it is obvious that concepts whose extensions are empty (even concepts that can be proved by logic alone to be empty) have a very important role in his science. Furthermore, since it is not always evident that a concept is contradictory, it may sometimes be of great importance to prove that a particular concept is contradictory. If we could not form contradictory concepts, none of this would be possible.

Perhaps this response seems to be a way of ducking the question. After all, one might suppose that Frege must recognize the existence not only of primitive concept-words but also of primitive object-names. Further, one might suppose that, if there are primitive object-names, it is permissible to introduce a primitive object-name for a fairy. Neither of these suppositions is warranted, however. It is, first of all, not obvious that there are primitive object-names.[6] The primitive terms, in both versions of Frege's Begriffsschrift, include no object-names. Nor are there any primitive object-names in geometry. Of course, this may simply be a symptom of the generality of logic and geometry. After all, these are both a priori sciences. Given Frege's assumptions about the available sources of knowledge for a priori science, knowledge of fairies, if there is such a thing, will be empirical. It may turn out that for some empirical sciences the introduction of primitive object-names is necessary. What would be involved in introducing a primitive object-name?

Primitive terms are introduced via elucidation, that is, some sort of prose gloss that will make the Bedeutung of the primitive term clear. But this is not to say that anything goes. The purpose of elucidation is to eliminate ambiguity in the basic terms used in some scientific research. Frege says,

> It is this, therefore, that serves the purpose of mutual understanding among investigators, as well as of the communication of the science to others. We may relegate it to a propaedeutic. It has no place in the system of a science; in the latter, no conclusions are based on it. Someone who pursued research only by himself would not need it. The purpose of explications is a pragmatic one. (CP 300–301)

The introduction of a primitive object-name, say, 'Titania', can be successful only if it names some empirical object that can be studied by investigators—otherwise there is no purpose to such an introduction. Furthermore, we must have noninferential justification for some claims about Titania.

If a primitive object-name is introduced for purposes of setting up a science of elves and fairies, it must, in particular, be possible to investigate whether or not the object in question is a fairy. But if the notion of fairyhood is significantly related to our everyday notion, the object in

---

[6]E.-H. Kluge argues that there cannot be primitive object-names (*The Metaphysics of Gottlob Frege*, 84–99). This is based on an elaborate metaphysical argument that cannot be reproduced on the sort of interpretation I am advocating.

question will not be a fairy. What if the concept of fairyhood is itself primitive? It is important to note that the primitiveness of the notion of fairyhood does not allow the stipulation that a particular object is a fairy, any more than the primitiveness of the notion of point allows the stipulation that a particular object is a point. 'Titania is a fairy' cannot be taken as an axiom of the science of elves and fairies any more than 'The extension of the concept *not self-identical* is a point' can be taken as an axiom of geometry. We cannot introduce 'Titania' with something like, 'The fairy who . . .' unless it can be shown or known in some other way that there is one and only one fairy who . . .

A fairy must be a magical being (otherwise the science in question would not be a science of elves and fairies). In fact, if the meanings of 'Titania' and 'fairy' can be made clear and if fairyhood, on this explanation, is understood in anything like a traditional way, it should be possible to show that Titania is not a fairy. Just as the study of houses that—according to their occupants—are plagued by poltergeists can reveal that something less mysterious is responsible for the odd events that have been attributed to the action of poltergeists, a study of the object that has been named 'Titania' will, presumably, show that (perhaps contrary to initial appearances) this object is not a magical being. Indeed, this suggests that 'Titania' and 'fairy' cannot both be primitive terms. For, were they primitive terms, then either 'Titania is a fairy' or 'Titania is not a fairy' would be a primitive truth—a truth that could be justified noninferentially. But if the investigation of the fairyhood of Titania is like the investigation of the existence of poltergeists, it will involve inferences.

Once again, I may seem to have gone around in a circle. Earlier I said that Frege's notion of meaning does not involve some sort of hook to the world but only precision of expression. In the above discussion, however, I have suggested that for the introduction of an object-name what is needed is that we can come up with or investigate the object. But it seems that this amounts to saying that a hook to the world is exactly what is needed in order to attribute a Fregean meaning to an object-name. In order to address this issue, I will turn now to a discussion of the characterization of objectivity in chapter 2 of *Foundations*.

Frege says very little about objectivity in *Foundations*. On the other hand, in the late essay "Thoughts," not only does Frege explicitly talk about objectivity, he gives a recognizable argument against idealism.[7]

---

[7]See CP 365–367, where Frege argues that he himself is not an idea.

Nonetheless, I have chosen to discuss Frege's understanding of objectivity by concentrating on passages from *Foundations*. What is the rationale for this choice?

My aim is to look at how these particular views contribute to Frege's project of defining the numbers and proving the truths of arithmetic from the laws of logic along with definitions. This project began with Frege's *Begriffsschrift* and was to have been completed in his *Basic Laws*. "Thoughts" was published long after both volumes of *Basic Laws*. Thus, were there views that appear only in "Thoughts" and have no substantive connection with his earlier works, it would be unclear that they could be understood as playing a role in this project. A view about objectivity which plays a role in the conception of Frege's project must be a view that Frege held long before he wrote "Thoughts." This is my reason for concentrating on *Foundations* rather than "Thoughts." But "Thoughts" is not discontinuous with Frege's earlier works. As I read "Thoughts," it is a contribution to his battle against the psychological logicians, and the views expressed there are entirely consonant with those expressed in his earlier works. This is not to say that there is convincing evidence for taking everything Frege says in "Thoughts" as an expression of his earlier views. Rather, there is an outline of a view of objectivity in *Foundations*, and some of the details of this outline can be filled in by looking at passages from "Thoughts." In the following discussion of chapter 2 of *Foundations*, I will do just that.

After arguing, in chapter 1 of *Foundations*, that the numbers cannot be taken as primitive, Frege turns his attention, in chapter 2, to an examination of the general concept of number and, in particular, how it might be defined. As in chapter 1, he approaches his question through a discussion of the views of other writers. Also as in chapter 1, the purpose of his discussion is to lend more plausibility to his (initially implausible) view that arithmetical truth must be analytic. To this end Frege criticizes attempts of other writers to show that the concept of number is indefinable as well as attempts of other writers to define the concept of number.

Most of these criticisms of other writers are brief and dismissive. In the first three sections of chapter 2 of *Foundations*, which, together, occupy two and a half pages, Frege considers both geometrical (or synthetic a priori) definitions of number and the available arguments that the concept of number is indefinable. These sections complete Frege's consideration of the view that our understanding of the general concept of number is synthetic a priori. The final section of chapter 2, which occupies just under one page, contains objections to using the notion of set to define that of number. These objections, presumably, are needed in order to distinguish Frege's ultimate characterization of the concept of number in terms of extensions of concepts from set-theoretic characterizations. Given Frege's aim of making it plausible that arithmetical truth is analytic, the remaining task is to argue against taking the concept of number to be definable in terms that belong to the realm of the synthetic a posteriori, that is, against a definition that will make the truths of arithmetic into empirical truths. The bulk of chapter 2 is devoted to such an argument.

These sections, I will argue, contain Frege's substantive attack on empiricist accounts of arithmetical truth. This attack, however, is not limited to empiricist accounts of *arithmetical* truth. For implicit in it is Frege's alternative to empiricism as well as his critique of what he takes to be the fundamental assumption of empiricism.

Although Frege mentions not only Mill but also Locke and Berkeley in this chapter of *Foundations*, his attack on empiricism is not directed at any of these philosophers in particular. The only writings considered at any length are Mill's, and, as in the first chapter of *Foundations*, Mill's views are not really taken seriously. The attack is directed at any episte-mological view on which all objective knowledge must be dependent on sense impressions.

It is important to remember that, given Frege's understanding of epistemology, it does not follow that it is possible that we could actually make any judgements (correctly or otherwise) without sense impres-sions. In fact, after giving his own analysis of the content of number-words—an analysis on which the source of our knowledge of the truths of arithmetic is not sense experience and our knowledge of these truths is independent of sense impressions—Frege issues the following re-minder:

> By this I do not mean in the least to deny that without sense impressions we should be as stupid as stones, and should know nothing either of numbers or of anything else; but this psychological proposition is not of the slightest concern to us here. Because of the ever-present danger of confusing two fundamentally different questions, I make this point once more. (FA 115)

The fundamental difference is, of course, the difference between ques-tions that concern the psychological laws that determine our coming to make certain judgements about the numbers and those that concern the source of our knowledge (or ultimate grounds) of these judgements. It is consistent with Frege's view that it would be impossible for us to make any judgements without sense impressions yet that sense impressions do not count as part of the source of our knowledge of arithmetic.

Frege has, of course, already criticized empiricist accounts of arithmetical truth. Several sections of chapter 1 of *Foundations* were devoted to extensive criticism of passages from Mill's *System of Logic*. These arguments, however, are quite different in nature from the arguments that appear in chapter 2. The purpose of Frege's first chapter is to convince his readers that the numbers and the concept of number are not primitive and, consequently, must be definable. Thus, in his chapter 1 discussions of Mill, the only empiricist claims Frege considers and criticizes are straightforward accounts of the foundation of our knowledge of arithmetical truth—accounts that leave no room for analyses of the number one or the concept of number. Frege takes these criticisms to establish that the numerals and 'number' cannot be primitive terms of an empirical science.

Since the burden of the arguments in chapter 1 was to show that the terms belonging to arithmetic cannot be primitive, the criticisms from this chapter do not establish anything of significance about the sort of definition required for the concept of number, should it be definable. In order conclusively to show that arithmetical truth is not empirical, Frege must say something about this. For Frege begins chapter 2 with the claim that the general laws of arithmetic will be derivable from a definition of the general concept of number. If this is so and if the concept of number must be given an empirical definition, then the general laws of arithmetic will be empirical and it follows that all arithmetical truth is empirical. Although he is not explicit about this, Frege's real objection to empiricist accounts of arithmetic can be found in his discussion of what is required of a definition of the concept of number.

Frege's argument that the concept of number cannot be given a definition that will make the truths of arithmetic into empirical (or synthetic a priori) truths is divided into two parts headed "Is Number a property of external things?" and "Is number something subjective?" This division seems natural given Frege's characterization, in chapter 1, of empirical propositions as those that "hold good of what is physically or psychologically actual" (FA 20).[8] But how does this fit with the

---

[8]In view of Frege's later understanding of 'actuality' (*Wirklichkeit*) as, roughly, the ability to act (see CP 370–372), it may be important to note that the psychological actuality

characterization of synthetic a posteriori truths as truths whose ultimate ground is sense experience? For the naive reader whose acquaintance with Frege's work is confined to *Begriffsschrift* and *Foundations*, there is no obvious difficulty. The senses constitute the source of our knowledge of what is physically actual. The candidates for what is psychologically actual are, presumably, mental states or inner images. Frege says, "An idea in the subjective sense is what is governed by the psychological laws of association; it is of a sensible, pictorial character" (FA 37). The source of our knowledge of what is of sensible, pictorial character should, presumably, also be the senses. The source of our knowledge of psychological laws should also be the senses.

The problem with this characterization is that it appears to conflict with Frege's epistemological picture. The sources of knowledge are sources of objective knowledge. How could we have objective knowledge of subjective ideas? In fact, there is no claim in *Foundations* that we can have objective knowledge of subjective ideas. On Frege's view, we can have objective knowledge of psychological laws, not of the character of subjective ideas.[9] Thus, the characterization of empirical propositions as those that hold good of what is physically or psychologically actual is accurate, if somewhat misleading. The views on objectivity which can be found in chapter 2 should clear up any confusions that might arise from this characterization. There remains a puzzle. Why should Frege have introduced a misleading characterization? After all, the objective science of psychology does not seem to be appreciably different in its methods of justification from any other physical science.

---

mentioned in this passage is not the ability to affect mental processes. For, in Frege's later sense, thoughts act by affecting the mental processes of a thinker, but Frege surely does not mean to say, in *Foundations*, that all knowledge about thoughts is empirical. As Gregory Currie notes, in *Frege: An Introduction to His Philosophy* (194), Frege's use of this term is not always consistent.

[9]Actually, there is one point in "Thoughts" at which Frege seems to deny this. After contrasting the inner and external realms, Frege says, "And here I come up against a further difference between my inner world and the external world. I cannot doubt that I have a visual impression of green, but it is not so certain that I see a lime-leaf. So, contrary to widespread views, we find certainty in the inner world, while doubt never altogether leaves us in our excursion into the external world" (CP 367). It is important to notice, however, that this is not a claim that we can have objective knowledge about subjective ideas. Frege's certainty that he has a visual impression of green is not a certainty about the subjective character of his visual impression. Greenness is an external, not internal, property. As Frege says earlier in the same discussion (CP 364), there are no green ideas. Frege's certainty is a certainty about his mental state—a certainty that might well be cashed out by objective psychological laws.

Why explicitly distinguish psychology from other physical sciences in a characterization of what is empirical? The answer, I suspect, has to do with the necessity of continued vigilance against psychologistic interpretations of his logic. The characterization of empirical propositions is yet one further reminder that Frege's logical source of knowledge is not to be confused with a psychological source of knowledge. Logical truths, unlike psychological truths, require no support from the senses.

Frege's discussion in sections 21–27 of *Foundations* is designed to exhibit deficiencies in the available empiricist strategies for defining the terms of arithmetic. What resources are available if sentences in which number-words appear must express claims that hold (or fail to hold) of the empirical world?

Frege introduces the project of trying to define the concept of number with a discussion of the grammatical roles played by number-words. The first step in trying to define the concept of number is to determine whether number-words (e.g., 'one', 'two') are object-words or concept-words. The obvious strategy for distinguishing object- and concept-words is to examine the grammatical roles played by the terms in question. Concept-words generally play the grammatical role of adjectives, for instance. However, given Frege's requirements, this strategy does not make the nature of number words any clearer. Frege requires that the definitions allow us to replace everyday arithmetic with a systematic science of arithmetic. Since everyday arithmetic involves not only pure arithmetic but also assertions involving applications (in particular, such claims as 'Here are two horses,' which Frege calls "assertions of number"), the definitions must be suitable for explaining the justification of both sorts of assertions. Frege's definition of the number two must, for instance, be appropriate for use in explanations of the content and justification both of '1 + 1 = 2' and of 'Frege owned two horses'. Grammar is of no help here. For in the above sentences, '2' seems to play the grammatical role of an object-word, while 'two' seems to play the role of a concept-word. The assertion that $1 + 1 = 2$ seems to be that 1 and 2 stand in a certain relation to each other just as, in "Frege wrote *Foundations*," Frege and *Foundations* are said to stand in a certain relation to each other. On the other hand, 'two' appears to play the same grammatical role in 'Frege owned two horses' as 'red' plays in 'Frege owned red horses'. The conflict here indicates that this is one of the situations in which grammar can lead us astray. In order to give the definitions Frege requires, we must find a way to restate either the sentences of arithmetic or sentences used in the applications of arithmetic so that number-words are used in the same way in both.

It is important to see that this is *not* to say that number-words actually *are* used in the same way in both sentences of arithmetic and sentences of its applications. The conflict above indicates that they are not. But the fact that there is a conflict, that the grammar of these uses of number-words is different, indicates that grammar is not a good guide for exhibiting the logical role of number-words. Furthermore, since everyday science requires both uses of the number-words, any definition of the number-words must provide a means for justifying claims made by sentences involving either use. In order for a definition of a number-word as an object-word, for instance, to satisfy this requirement, it is necessary that what is expressed by sentences in which number-words appear as concept-words be restated. Similarly, a definition of a number-word as a concept-word will only satisfy the requirement if the sentences of pure arithmetic can be restated. But which sort of sentence should be restated? And what criteria should determine which should be restated?

For someone who thinks all knowledge must be empirical, the most reasonable strategy would seem to be to take assertions of number as basic and to try to restate propositions of pure arithmetic. After all, arithmetic does not *seem* to be about psychology, and since assertions of number (i.e., claims such as "Here are four horses") *do seem* to be about the physical world, the first strategy for giving an empiricist account of arithmetic might be to understand propositions of arithmetic as facts about physical (or sensible) objects. But 1 and 2 are not physical or sensible. Thus, the claim that $1 + 1 = 2$ must be restated if it is unambiguously to express a fact about (or generalization of facts about) physical objects. Consequently, it seems only reasonable to take assertions of number, rather than arithmetical propositions, as basic. Since numbers are not physical objects, there seems to be no alternative but to take numbers to be properties of physical objects. Frege begins the substantive part of chapter 2 of *Foundations* with a criticism of this first-strategy empiricist view.

The first-strategy view is represented, for Frege, by M. Cantor and E. Schröder. The Schröder view, in particular, has pretensions to giving an account of at least some of our knowledge. Frege says:

> For E. Schröder number is modeled on actuality, derived from it by a process of copying the actual units with ones, which he calls the abstraction of number. In this copying, the units are only represented in point of their frequency, all other properties of the things concerned, such as their colour or shape, being disregarded. . . . It follows, therefore, that Schrö-

> der puts frequency of Number on a level with colour and shape, and
> treats it as a property of things. (FA 27–28)

This view cannot, of itself, answer Frege's questions. First, in order to
extract the proper sort of account of knowledge from this view, there
must be recognizable units in actuality, and we must be able to give
some kind of account of how it is that we are presented with these units.
However, Frege argues, it is not possible even to satisfy the first of these
necessary conditions. Frege attributes the decisive objection to Bau-
mann, who argues that it is simply false that we are presented with
strict units; the isolated groups or sensible points with which we are
presented can each be viewed either as single units or as many units.
This is a central argument for Frege, and he elaborates on it at some
length.

Frege takes the *Iliad* as an example. With what units does the *Iliad*
present us? The problem with answering this question is that the answer seems to depend on how we think of the *Iliad*. If we are thinking of
poems, we seem to be presented with one and only one unit. However,
if we are thinking of books, we seem to be presented with twenty-four
entirely different units. If we think of verses or words, we are presented
with still more entirely different units. The *Iliad* itself does not seem to
present us with units—what counts as a unit seems to depend on how
we are thinking of the *Iliad*. Frege goes on to point out that it seems to
follow that various distinct numbers can be ascribed to the same physical objects. What is the significance of this?

Consider some physical property, say, color. Ideally, for every color,
every object either is or is not that color. That is, if we understand that
an object has a given color just in case its entire surface area has the
power of reflecting light of certain wavelengths and absorbing to various extents light of other wavelengths, then every object either is or is
not a particular color.[10] In this sense, every chair either is green or is not

---

[10]Of course, many ascriptions of color are vague and problematic. It is not at all clear, for
instance, where to draw the boundary between red and purple. Thus, one might want to
claim that there are some borderline cases—some objects that cannot truly be said to be
red or purple and hence are neither red nor not-red. However, it is a consequence of
Frege's view of the logical laws as applying universally (an instance, given this universality, of the law of the excluded middle) that such an object cannot be neither red nor not-
red. This may well seem to present a serious problem for Frege. It is, however, distinct
from the problem with number raised here. For, while there are some unproblematic
cases for color-words (e.g., some clearly red objects), I will go on to argue that, on Frege's
view, there are none for number-words (e.g., no objects that are clearly four). Frege's

green—a chair that has both red parts and green parts is not green. Similarly, it is unproblematic to say that every idea either is or is not green. Since no ideas have surfaces, no ideas are green.[11] It is true of every idea that it is not red, not white, not black, and so on. In this cleaned-up sense of color, to say that an object is (is not) green (red, black, white) is always to make a substantive claim.

But, as the *Iliad* argument shows, these features of colors do not apply to numbers. The claim that a particular chair, for instance, is twenty-four is nonsensical. One would have to say that the chair is twenty-four somethings. It might, of course, be that the chair is made of twenty-four wooden planks. And, if it is understood in advance that the chair is to be viewed as wooden planks, then one might be able to say, as an abbreviation, that the chair is twenty-four. However, such a claim is clearly an abbreviation of the claim that the chair is twenty-four wooden planks. This is a clear indication that it cannot be the *chair* that has number. Indeed, if the chair is made of six interlocking modular parts (each composed of four of the wooden planks), it would be as accurate to say that the chair is six as that it is twenty-four. But if the chair is six, it is less than twenty-four—hence the chair is not

---

point is that, because of this difference, it is impossible to treat number-words in the same way as words that represent physical properties. Furthermore, as a result of this difference, the vagueness problem cannot even be articulated for number-words qua physical property–words. It is, in any case, an assumption of Frege's that, if there are color concepts, these concepts have sharp boundaries. The problems introduced by the inadmissibility of vague predicates will be discussed in chapter 5.

[11]In fact, in a discussion from "Thoughts," Frege says that there are no green ideas (CP 364). This issue may seem unimportant. After all, we are only interested in making color ascriptions to physical objects. It may seem inappropriate even to discuss whether or not ideas are green. However, as I pointed out in chapter 3, it is a central feature of Frege's logic that every concept must either be true of, or false of, each object. The sun, for instance, must be either even or odd. Hence, if there is a concept of greenness, it must be possible to say whether or not ideas are green.

But Frege does say something in *Foundations* which may, at first glance, appear to contradict this: "An a priori error is thus as complete a nonsense [*Unding*] as, say, a blue concept" (FA 3). It may seem to be a consequence that he was in error in "Thoughts" to say there are no green ideas, and that his view that concepts must have universal domain is mistaken. On closer inspection, however, it becomes clear that the *Foundations* claim licenses neither of these consequences. Frege's point is that there are no a priori errors— to say that a judgement is a priori is to make a claim about its justification, and if it has a justification, it is not an error. The nonsense is not the content of the expression 'a priori error' or 'blue concept', but rather a particular a priori error or blue concept. For it would be nonsensical to say that a judgement can be termed an a priori error or to say that a particular concept is blue. In fact, Frege's discussion, far from being an attempt to outlaw such expressions, illustrates their utility.

twenty-four. Thus, we are committed to the contradictory (hence unacceptable) consequence that the chair both is and is not twenty-four.

In fact, it is easy enough to draw even more absurd consequences. For, while it may seem natural to view the chair as made up of wooden slats, it is just as accurate to view it as made up of chair-halves. By reasoning as in the above case, it seems to follow that the chair is two (chair-halves) and three (chair-thirds)—hence not two. It follows that every object is both two and not two (three and not three, etc.).[12] It follows that the empiricist's most obvious strategy will not work.

The problem discussed above, of course, is simply a version of the *Iliad* argument, and the first-strategy empiricist view, the view that numbers are simple properties of objects, seems vulnerable to it. Our reluctance to license sentences like 'This chair is twenty-four' is not simply a reaction to linguistic awkwardness. Something is missing. An object-name and a number-word are not sufficient to express an assertion of number. An expression of an assertion of number must include something that indicates what is to count as a unit. Since physical objects do not present us with units, what counts as a unit will depend on how we have chosen to regard a physical phenomenon. But a choice of how to regard a physical phenomenon is a mental act. The upshot of the *Iliad* objection seems to be that assertions of number do not describe phenomena but, rather, certain sorts of mental acts, ways of regarding physical phenomena. It seems to follow that assertions of number belong to psychology. Of course, the above discussion begins with the assumption that assertions of number can be construed as simple predications. It may be that the conclusion that assertions of number belong to psychology can be avoided by another empiricist strategy.

In section 23 of *Foundations* Frege considers a more sophisticated

---

[12]Of course, there are other ways of looking at it. The claim that the chair is two chair-halves depends on our assuming some particular dividing line between the chair-halves. However, since there are uncountably many dividing lines between chair-halves, one might want to say that the chair is made up of uncountably many chair-halves. Similarly, the chair will be made up of uncountably many chair-thirds, and so on. On this way of looking at the situation, the upshot is that the chair (in fact, any object) does have a particular number, the number of the continuum. But this is as absurd as the other conclusion. If we are committed to the view that every object has the number of the continuum, then, although we can avoid the above contradictions, all previously accepted number statements must be false. The content of such claims as 'Here are two horses' will be entirely swallowed up by the fact that each horse has the number of the continuum; hence so does the number of the two horses together. Such an explanation of numbers will not meet Frege's requirement that the explanation allow us to make the content of everyday number assertions clear.

empiricist strategy, a strategy he attributes to Mill. Rather than claiming that numbers are properties of all individual physical objects, Mill says that numbers can be properties only of a very special sort of physical object—an agglomeration.[13] In fact, Mill provides something that looks like a definition of the concept of number. He says that a number is the characteristic manner in which an agglomeration is made up of, and may be separated into, parts (*System of Logic*, bk. 3, cap. xxiv, section 5). This may seem to provide a solution to the problem outlined above. For the notion of agglomeration seems to include in it a specification of unithood. Consider a pile of apples. The most natural way of separating this pile into parts would be to separate it into individual apples. If this is the characteristic manner in which the agglomeration of apples can be separated into parts, then the units are apples and the agglomeration *does* have a determinate number—it will be the number of apples.

This strategy seems to avoid the indeterminacy-of-number problem in an attractive way. For an obvious problem with the more absurd consequences noted above is that it is quite difficult to imagine circumstances in which the chair would be viewed as made up of chair-halves; such a division is markedly unnatural. On Mill's view, the claim that the chair is two chair-halves, three chair-thirds, and so on does not show that the chair is both two and not two. For, first of all, the chair will only have a number if it is an agglomeration. Presumably, it is not an agglomeration, has no number, and, consequently, is not two. No contradiction should arise.[14] But a commitment to empiricism, as Frege understands it, will ensure the failure of this strategy as well.

---

[13]Donald Gillies defends a similar view in chapter 5 of *Frege, Dedekind, and Peano on the Foundations of Arithmetic*. Gillies talks about naturally occurring aggregates or sets and suggests that, for mathematical reasons, it is more appropriate to found arithmetic on sets than on concepts. In light of the contradiction in Frege's system it may seem that the latter consideration is correct. For, as Frege himself came to recognize, arithmetic cannot be founded on concepts, and it is, of course, possible to give set-theoretic definitions of the integers. But the arguments below, on Frege's view, preclude founding arithmetic on the notion of naturally occurring aggregates, and the abstract notion of set is, on Frege's view, too vague to provide a foundation for arithmetic. The sort of foundation Gillies discusses does not answer to Frege's epistemological needs.

[14]Of course, the chair might be an agglomeration, in which case it would have a number. And it seems that, on Mill's view, if the chair is an agglomeration, then there must be a determinate sort of object of which the chair is an agglomeration. If the chair is an agglomeration of wooden planks, then its number is twenty-four; if it is an agglomeration of modular parts, its number is six; if it is an agglomeration of chairs, its number is one. Thus, if Mill is to avoid the Iliad objection, there will have to be correct and incorrect answers to the question of what the objects of which the chair is an agglomeration are. It may seem most accurate to say that the chair is an agglomeration of one chair—that the

In order to understand the problem with Mill's account, it is important to be clear about several details. For this strategy to provide a way out of the *Iliad* objection, numbers must be *physical* properties. Hence, if numbers are properties of agglomerations, agglomerations must be physical objects. And it must be a determinate *physical* fact that a particular number is or is not a property of a particular physical object (i.e., agglomeration). In other words, if there is an agglomeration with the property three and an agglomeration with the property four, they must be distinct—there must be a physical difference. Finally, this account of the concept of number must be accurate to the concept of number as we antecedently use and understand it. The most natural candidate for the sort of physical object which is an agglomeration would seem to be something like a heap or cluster of objects, where physical proximity determines membership in the heap or cluster. Once 'agglomeration' is understood in this way, however, Frege can level either of two fundamental objections to Mill's view.

The simpler of these objections directly concerns the interpretation of agglomerations as physical heaps. If, on Mill's account, numbers can be properties only of physical heaps, this account cannot be accurate to our concept of number. For many things that are not parts of physical heaps or clusters are actually counted. It is possible, Frege says, to determine the number of blind in Germany without ever bringing them together physically. It is also possible to count many things that could not be parts of physical heaps. Ideas, concepts, and proofs can be counted, but they cannot be piled together into heaps. Thus, if numbers can be properties only of agglomerations, agglomerations cannot simply be physical heaps. Furthermore, this is not a conclusion whose force is restricted to views on which agglomerations are understood to be physical heaps. For such things as ideas and concepts cannot ever constitute parts of physical objects. Thus, if numbers were to be properties of agglomerations, agglomerations could not be physical objects.

It may seem that Mill could respond by understanding agglomera-

---

chair is one. But if the chair is an agglomeration of chairs, then it cannot be an agglomeration of modular parts or of wooden planks. For this reason, it may appear that the most appealing solution is to say that the chair is not an agglomeration at all. It is not clear, however, that this will work. For, if Mill is to answer Frege's questions, he must have an account of why it is correct to say that there are twenty-four wooden planks here and that there is one chair. The twenty-four wooden planks must make up an agglomeration, and it will be difficult to find a physical difference between the agglomeration of planks and that of the chair.

tions not exactly as physical heaps but as conjunctiva of physical objects. That is, one might claim that there is an agglomeration of the telephone on my desk, the sun, and the moon. Similarly, one could claim that there is an agglomeration of the blind in Germany. This would circumvent the problem of our apparent ability to determine the number of blind in Germany without holding a rally. Furthermore, we do want to be in the position of being able to claim that the telephone on my desk, the sun, and the moon constitute three objects. Thus, some move of this nature seems only appropriate. However, Frege's other objection, which is a version of the *Iliad* objection, can be applied not only to agglomerations that are physical heaps but also to conjunctiva of physical objects.

Frege introduces this version of the *Iliad* objection by considering particular agglomerations, one of which is a heap of straws. In essence the objection comes to this. There is no physical difference between the heap, or agglomeration, of straws and the agglomeration of pairs of straws in the heap. Suppose there are 100 straws in the heap. Since this is an ordinary, informative use of a number-word, it must be possible to use Mill's definition to translate this claim into agglomeration-talk. This is not difficult. Mill need only say that the characteristic manner in which the agglomeration of straws can be separated into parts is 100. And this seems right. But agglomerations must be physical objects, and there is no physical difference between this agglomeration and the agglomeration of pairs of straws. It must follow, then, that the characteristic manner in which the agglomeration of pairs of straws can be separated into parts is 100. But Mill's definition must also account for the fact that we can, informatively and correctly, say that there are 50 pairs of straws. It must also follow, then, that the characteristic manner in which the agglomeration of pairs of straws can be separated into parts is 50; hence is less than 100; hence is not 100. That is, the characteristic manner in which this agglomeration can be separated into parts is not 100. This is a contradictory conclusion. For Mill's definitions to work, a particular agglomeration must have only one number. So the definition fails.

It is important to note that this argument is not dependent on taking the agglomeration to be a physical heap. No use was made of the role of physical proximity in determining membership in the heap. The argument turned only on the observation that there is no physical difference between an agglomeration (heap) of individual straws and the agglomeration (heap) of pairs of those straws. This argument will be applicable to any account on which numbers are viewed as properties of physical objects. For instance, if the conjunctive object that is the agglomeration of pencils in Germany is a physical object, it must be the same physical object as the agglomeration of pairs of pencils in Germany. The number of pencils in Germany must, of course, be distinct from the number of pairs of pencils. Thus, a view on which numbers are properties of physical objects cannot account for the justification of our everyday assertions of number.

This argument, by itself, does not provide a refutation either of empiricist accounts of arithmetical knowledge or of all accounts on which numbers are physical properties. It is instructive to look at exactly what assumptions are involved in the view Frege is criticizing. Frege's basic assumption, as always, is that the views discussed must answer his questions. That is, the view of numbers as properties of agglomerations must account for the truth of all those sentences in which number words appear and which we normally take to express truths. The particular assumptions involved in the view Frege is considering are that numbers are physical properties, and that only certain physical objects (agglomerations) possess these properties. The difficulty with these assumptions is that our ordinary assertions of number allow us to assign different numbers when there is no physical difference—the same physical object is both an agglomeration of straws and an agglomeration of pairs of straws. It seems to follow that assertions of number cannot amount to the assertion that a particular sort of object (an agglomeration) has a particular physical property (a number). The problem is that the identification of a physical object as an agglomeration does not suffice to specify unithood. An agglomeration of one sort of unit is equally, for instance, an agglomeration of pairs of that unit. As before, what appears to be missing is a choice, a mental act.

This is as far as Frege pushes this particular argument. In the next section, section 24, Frege gives more general reasons for not taking numbers to be physical properties. Before I turn to the discussion in section 24, however, a few comments about the above argument are in order. This argument is not, in itself, an objection to empiricism. The central point is that although any assertion of number depends on a specification of what is to count as a unit, the physical world does not present us with units (or, unithood is not a physical property). It appears that the upshot of this argument is that the determination of units and hence part of the justification of assertions of number must be psychological. Frege goes on to argue that psychology, in an objective sense, cannot figure in this justification; hence, for the empiricist, arithmetic must be subjective. This argument seems to trade on the contrast between assertions of number and assertions of everyday sensible properties. For the difficulties concerning unithood do not seem to arise in the latter case. In fact, however, Frege ultimately suggests that there are similar difficulties with the empiricists' view of apparently unproblematic claims about the physical world.

It may seem that the initial part of Frege's argument is flawed by his omission of the more plausible empiricist strategies. In particular, he only considers views in which numbers are physical properties either of objects in general or of particular sorts of objects (agglomerations). In these cases, what is missing from any attempt to make an assertion of number is the specification of unithood. But there may seem to be more sophisticated empiricist views on which unithood is specified. Suppose that instead of taking numbers to be physical properties, numbers are taken to be complicated entities involving both a physical object and a part-whole analysis.[15] It might seem that, if part-whole analyses can be viewed as unproblematic physical relations on the order of, say, relations of comparative height, Frege's objection could not get started. But it is no clearer that the physical world presents us with part-whole analyses than that it presents us with units. Frege's *Iliad* argument can be rephrased as an argument that physical objects do not present us with a particular part-whole analysis any more than they present us with units.

Thus, I suspect that Frege would be as willing to contrast part-whole analyses and physical relations as he was to contrast numbers-as-properties and colors. Indeed, as I will indicate below, the contrasts drawn between colors and numbers-as-properties in section 24 apply as well to physical relations and part-whole analyses. But since Frege did not explicitly consider part-whole analyses, I do not want to attribute any views about these relations to him. In any case, it is not important to attribute any views on this issue to Frege because, ultimately, the contrast Frege draws is only a device for illustrating the necessity of some knowledge that cannot be obtained through the senses for the justification of assertions of number. But this contrast is not meant to hold up. On Frege's view, empiricists are as doomed to admitting some source of knowledge other than the senses as part of the justification of

---

[15]I am indebted to an anonymous reviewer for this suggestion.

assertions of color as they are to admitting such a source as part of the justification of assertions of number.

In the next section, section 24, Frege gives more general reasons for not taking numbers to be physical properties. The problem with identifying numbers with physical properties is not simply that physical differences need not accompany assertions of different numbers. For it is also true that correct assertions of number can be made which involve no physical objects at all. Ideas, events, and judgements can be counted, but none of these is a physical object. Frege believes that it simply makes no sense to attribute physical, or sensible, properties to nonphysical objects. Such considerations, of course, apply equally to part-whole analyses. For such analyses apply to ideas, events, and judgements as well as to physical objects.

Frege's discussions in this section are meant less as substantive arguments against positions people have actually held than as illustrations of how absurd it would be for anyone to attempt to support such a position. After all, if numbers are just ordinary physical properties like tastes and odors, how can it be that, while we cannot taste or smell ideas and hence cannot correctly ascribe flavors or odors to them, we can correctly assign numbers to them? Frege also suggests that our ability to ascribe a particular physical property to objects is dependent on our having a sensory experience of an object with that property and on our being able to recognize some feature of that experience as corresponding to the property.[16] But, he suggests, it would be absurd to say that something nonsensible can have sensible properties. Thus, it is really impossible to give a plausible account of the justification of the claim, say, that the number of figures in a syllogism is four. Since syllogistic figures are not physical objects, fourness cannot be a feature of our experience of them. One might want to claim that fourness is a feature of our experience of the symbols for the syllogistic figures. But it is not at all clear that this will help justify the claim that the number of figures is four. For there are many situations in which distinct symbols stand for the same thing. It would be legitimate to infer that the number of figures

---

[16]It should be noted that Frege does not really believe this. For he also suggests, in section 26, that a colorblind person can still correctly ascribe redness to certain objects. However, the colorblind person's ability to do this is dependent on the ability of other people to experience redness. In the number case discussed in section 24, the point is that *no one* can experience, say, threeness. There is no way for us correctly to ascribe a physical property to some object if that property can be experienced by no one. To deny this would be to deny the claim that our knowledge of the physical world is synthetic a posteriori.

is four only if it could be shown that there were the same number of symbols as of figures. And, Frege notes, there is no reason for the number of symbols to coincide with the number of things symbolized.

These arguments complete the first part of Frege's attack on the empiricist explanation of numbers. Numbers can be neither physical objects nor properties of physical objects. But, more significant, Frege has argued that the physical world does not, on its own, present us with everything necessary to justify an assertion of number.[17] For the empiricist, the only other remaining strategy is to construe assertions of number as involving not only what is physically actual but also what is psychologically actual. It does not follow immediately that the empiricist must take arithmetic to be subjective. For even Frege believes that there are objective psychological laws; among these are the laws in accordance with which we do judge as opposed to the (logical) laws in accordance with which we should judge and the laws of associations of ideas. Thus, on Frege's view, the empiricist might attempt to base arithmetic on psychology in this objective sense.

While this move may initially have seemed unappealing—as Frege says, "It would be strange if the most exact of all sciences had to seek support from psychology, which is still feeling its way none too surely" (FA 38)—Frege's arguments against taking numbers to be physical properties have made this move much more attractive. For most of Frege's arguments against Mill consist of discussions of cases in which there are differences in number which correspond to no physical differences. The discussion of the heap of straws which is both 100 straws and 50 pairs of straws seems to count as a straightforward argument that assertions of number do not express physical facts. There is no physical difference between the heap of straws and the heap of pairs of straws. The only difference seems to be the manner in which the heap is regarded. If one is allowed to regard agglomerations as nonphysical, it may well seem that Mill is on the right track, that agglomerations do have numbers as properties. After all, there seems to be nothing wrong, in principle, with counting the moon, the sun, and my telephone. One might say that these objects make up an agglomeration by virtue of our

---

[17]In a discussion of the empiricist's suspicion of abstract objects in *Frege's Conception of Numbers as Objects*, Crispin Wright attributes a similar argument to Frege. Wright views some of the empiricist's suspicion as a result of taking ostensive definition as a model for determining the reference of a singular term, and he offers a Fregean argument that ostensive definition can work only against a "fairly elaborate background conceptual equipment" (44).

regarding them as making up an agglomeration. Frege brings such considerations up explicitly in section 25.

Frege's discussion begins with a restatement of his earlier arguments against taking numbers to be physical properties. Oneness or twoness, he says, surely could not be physical, else *one* pair of boots could not be the same physical phenomenon as *two* boots. What, then, can the empiricist identify as oneness or twoness? While there are no physical differences in this example which correspond to the different numbers ascribed, there *is* a difference, which may be acknowledged by the empiricist, which corresponds to the difference in number. This is a difference in the psychological states of the person who judges that there is one pair of boots and the person who judges that there are two boots. Thus, it is plausible that there may be a sense in which the laws of number can be viewed as psychological laws. Basing arithmetic on psychology no longer looks like such a desperate move.

Frege seems to regard it as an immediate consequence that the empiricist is committed to the subjectivity of numbers. His discussion of the absence of physical differences between one pair of boots and two boots ends with the following quotation from Berkeley:

> It ought to be considered that number . . . is nothing fixed and settled, really existing in things themselves. It is entirely the creature of the mind, considering, either an idea by itself, or any combination of ideas to which it gives one name, and so makes it pass for a unit. According as the mind variously combines its ideas, the unit varies; and as the unit, so the number, which is only a collection of units, doth also vary. We call a window one, and yet a house in which there are many windows, and many chimneys, hath an equal right to be called one. (FA 33)

Section 25 ends with this quotation; the next two sections are introduced with the heading "Is number something subjective?"

Frege assumes throughout his writings that there are objective truths of arithmetic. His aim in *Foundations* is to make it plausible that the source of our knowledge of these objective truths is the logical source. To this end, he wants to exhibit the unsatisfactory nature of other accounts of the source of our knowledge of these truths. He does not, however, need to argue that there are objective truths of arithmetic. If empiricism forces one to regard what look to be sentences of arithmetic as not expressing objective truths, Frege is free to disregard it. This view of the sentences of arithmetic, however, is not an immediate consequence of taking numbers to be subjective ideas. Even if numbers are subjective ideas, there may appear to be an option for the empiricist who wants to maintain that there are objective truths of arithmetic. This is to take the truths of arithmetic either to be or to require for their justification objective psychological truths. The problem with taking this option is that it leaves the empiricist no means for answering Frege's questions about the justification of arithmetical truths. In fact, given the constraints Frege places on an account of our the truths of arithmetic, this move commits one to the view that arithmetic is subjective. The arguments are closely related to Frege's arguments against psychologism.

Consider how objective psychological truths might be involved in the justification of the claim 'Here are two packs of playing cards'. Frege's earlier arguments indicate that the mere physical phenomenon, along with an idea of twoness, is insufficient to justify such a claim. Something else is needed. What could this be? Among the causes which give rise to the judgement that there are two packs of playing cards are antecedent decisions. Frege says, "The Number 1, on the other hand, or 100 or any other Number, cannot be said to belong to the pile of playing cards in its own right, but at most to belong to it in view of the way in which we have chosen to regard it" (FA 29). One antecedent decision, then, is to regard the pile as a pile of packs of cards rather than as a pile of cards. But this decision is not sufficient. For Frege continues:

> What we choose to call a complete pack is obviously an arbitrary decision, in which the pile of playing cards has no say. But it is when we examine the pile in the light of this decision, that we discover perhaps that we can call it two complete packs. (FA 29)

Thus, in order to judge whether a physical phenomenon is two, we need antecedently to decide how to regard that physical phenomenon, and this decision is a matter of arbitrary choice—the physical phenomenon does not determine this decision. How could such decisions play a role in the justification of the claim that there are two packs of playing cards in a particular pile?

Since the claim in question cannot qualify as a primitive truth, its justification will be inferential. If such decisions are to play a role in the justification, they will have to do this in propositional guise. That is, the claims 'I have decided to regard this phenomenon as consisting of complete packs' and 'I have decided to understand a complete pack as consisting of . . .' must be part of the justification of 'Here are two packs of playing cards'. But are these decisions necessary for the justification of 'Here are two packs of playing cards'? Suppose it is true that there are two complete packs of playing cards in the pile. Now suppose I were to decide to regard the phenomenon as consisting of cards. It would not follow that there were not two complete packs of playing cards in the pile. Rather, my judgement in this circumstance would not concern the truth or falsity of the claim that there are two packs of playing cards in the pile. Nonetheless, it would still be true that there were two packs of playing cards in the pile.

The decision about what to regard as constituting a complete pack may seem more relevant to justification but, in fact, its status is the same as the other decision. Were I to decide, for instance, that any two playing cards constituted a complete pack, I would then deny 'There are exactly two complete packs in this pile'. But my denial would not contradict the original claim 'There are two complete packs in this pile', for the result of my decision is that I am using the sentence to express something different from what the sentence was originally used to express. It is not the sentence that is denied but what the sentence expresses. As Frege says in "Thoughts," a thought (what a sentence expresses) can only be presented "wrapped up in a perceptible linguistic form" (CP 360); thus, he is sometimes compelled to talk about language when his concern is actually not with language at all.

If these decisions about how to regard a phenomenon or how to understand the terms used to express a judgement are not part of its justification, what is the relation between these decisions and the judgement? It may well be that it is psychologically necessary to make these decisions in order to judge that there are two complete packs of playing cards. For the very act of formulating the content of the judgement involves regarding the physical phenomenon (whether accurately or inaccurately) as consisting of packs of playing cards. The decisions involved are, as Frege indicates (FA 29, 34), completely arbitrary. Nothing prevents us, for instance, from regarding the phenomenon as consisting of horses and judging, correctly, that there are no horses. Alternatively, nothing prevents us from using the expression 'complete pack of playing cards' to express what we generally use the term 'horses' to express and, consequently, from asserting correctly, 'There are no complete packs of playing cards here'.

These decisions concern, in Frege's words, the question "of how we arrive at the content of a judgement" (FA 3). Furthermore, the purpose of Frege's introduction of the question of how we arrive at the content of a judgement is to distinguish this from the question with which Frege is concerned: "Whence do we derive the justification for its assertion?" The above considerations illustrate this distinction. From Frege's perspective, it is impossible for objective psychological laws to play a role in any justification other than that of psychological claims. Although the arrival at the content of these judgements must be psychological, the content itself is not psychological. Once we have arrived at the content of these judgements, no psychological decisions or considerations play a role in its justification.

Thus, given Frege's understanding of justification, the empiricist can only base arithmetic on psychology by resorting to the claim that arithmetic is subjective after all. Once the empiricist who has undertaken the burden of answering Frege's questions abandons the view that numbers are physical objects or properties, she or he is committed to taking arithmetic to be subjective.

Of course, once the empiricist takes arithmetic to be subjective, it will be impossible for the empiricist to answer any of Frege's questions. If numbers are ideas or psychological states, then each number must be *some person's* idea or psychological state. He says, "We should then have it might be many millions of twos on our hands. We should have to speak of my two and your two or one two and all twos" (FA 37). There can be no question of the justification, for instance, of 2 + 2 = 4, for it would depend on *whose* ideas of 2 and 4 are under consideration. In this case, we cannot talk about general, objective, psychological laws, we can only talk about individual, subjective, ideas or psychological states.

Frege's argument, so far, is as follows. He began with an argument that the empiricist cannot give an account on which assertions of number are purely physical facts. That is, numbers are neither physical objects nor sensible properties and, furthermore, the evidence available from our mere experience of a physical phenomenon is insufficient to justify the judgement that a particular number is a correct answer to the question "How many?" Frege's point is that something in addition to physical phenomena and numbers must be involved in the justification of assertions of number. The only possibility for the empiricist, on Frege's view, is that assertions of number must, in some way, be dependent on psychological facts. But objective psychological laws cannot enter into the justification of truths of arithmetic (or assertions of number). Hence, the empiricist is committed to taking the truths of arithmetic to be subjective. Although on Frege's view such a position is simply wrong and this shows that the empiricist cannot account for our knowledge of the truths of arithmetic, it is not obvious that Frege's empiricist need regard this as a reductio of empiricism. In fact, Frege's quotation of Berkeley seems to indicate that some empiricists might want to regard arithmetic as subjective. The next part of Frege's argument, however, may be considerably more worrisome. For Frege goes on to argue that empiricism entails not only that arithmetic is subjective, but that everything is.

Frege's argument begins with a comparison of the numbers to the North Sea. It seemed, from the arguments in the last section, that numbers might be creatures of the mind because the number of things, say, in a box will depend on an arbitrary choice (for instance, whether to regard the contents of the box as boots, pair[s], or even molecules). But, Frege argues in section 26, a similar arbitrary choice is necessary for answering any questions about the North Sea. Marking off a particular part of the water on the earth's surface and calling this the 'North Sea' was certainly a matter of an arbitrary choice to regard a particular body of water in a particular way. Of course, once the choice of boundaries for what will be called the 'North Sea' has been made, what lies within the boundaries of the North Sea is fixed and objective. Similarly, once the choice of whether to regard a phenomenon as boots or pair(s) has been made, the number is as fixed and objective. In other words, if the concept of number is subjective, so is the North Sea.

It may seem, however, that we are free to change our way of regarding the boots (and thus the correct number) in a way that does not apply to the North Sea. But this is not obviously true. There certainly could be circumstances under which we might choose to change the boundaries of what is called the 'North Sea'. So, once more, if it is legitimate to infer, from the fact that the number assigned to a particular phenomenon varies according to how we view it, that the concept of number is subjective, then similar reasoning should show us that the North Sea is subjective. For this reason the empiricist seems to be committed to admitting that the North Sea is subjective. But the North Sea is not subjective. The real conclusion that, Frege says, must be drawn is that this

> number is no whit more an object of psychology or a product of mental processes than, let us say, the North Sea is. The objectivity of the North Sea is not affected by the fact that it is a matter of our arbitrary choice which part of all the water on the earth's surface we mark off and elect to call the "North Sea." This is no reason for deciding to investigate the North Sea by psychological methods. In the same way, number, too is something objective. (FA 34)

Frege has thus offered what may be taken as a reductio of empiricism. In so doing he takes a risk. For, in spite of the above denial, it is not clear that Frege's argument does not, ultimately, amount to a straightforward defense of the claim that arithmetic, as well as everything else, is subjective. It now seems that the burden is on Frege to offer an alternative account of the justification of our knowledge which will enable him to vindicate the claim that arithmetic is objective. To see where his views on objectivity come from, it will be useful to say a little more about the problem with empiricism.

It is important to recall that Frege's central point is that something other than the mere perception of physical phenomena must be involved in the justification of our judgements. If one is committed, as Frege assumes the empiricist is, to the view that what is objective is what is physically or psychologically actual, then what is missing must be psychological. If all justification is based on psychology, Frege argues, there is no way for us to have objective knowledge. But Frege is not forced to conclude that what is missing is psychological. For Frege what is missing is logical. As he says, in *Foundations*, "Observation itself already includes within it a logical activity" (FA 99). What, exactly, is the missing logical part of the justification of our everyday assertions?

In each of the above examples, the truth of the assertion in question seemed to depend on arbitrary choices. Among the arbitrary choices involved in the cases of the phenomena involving boots and cards were choices of whether to regard the phenomena as consisting of boots or pairs of boots, cards or packs of cards, and how to understand 'pack of cards'. In the North Sea example the arbitrary choice had to do with drawing the boundaries of what is to count as the North Sea. But the psychological activities of making these choices were not themselves part of the justification of the assertions in question. Rather, the part of the justification which is missing from the empiricist's account is the result of this psychological activity, the boundaries that were drawn. What is needed for justification are concepts and, in many cases, their definitions.

Sense experience on its own will not enable someone to judge, for instance, that the North Sea is 10,000 square miles in extent. For such a judgement it is necessary to know what the borders of the North Sea are and what it is to be 10,000 square miles in extent. Another way of saying this, in Frege's later vocabulary, is to say that what is necessary is that the thought expressed by 'The North Sea is 10,000 square miles in extent' must be grasped. Similarly, the sensory experience that allows someone to claim that the petals of a flower are of some particular color is useless without the knowledge necessary for understanding what it is to be a petal of a flower. The problem with basing knowledge on the senses is not that the senses are so easily deceived but, rather, that the senses simply cannot give us knowledge on their own. Without concepts, sensory experience will be unintelligible and subjective—a sort of "raw feel." But, since concepts are not parts of the physical world, for the empiricist concepts must be part of the inner world. If the concepts of being a petal of a flower and of redness can only be parts of someone's inner world, then the judgement that a flower's petals are red can only be a judgement about some features of someone's inner world. And if each person's assertion that a flower's petals are red is an assertion about her or his inner world, there is no common content.

Frege claims that concepts are not part of the inner world. In a footnote (FA 37) he contrasts subjective ideas with ideas in the objective sense—which are nonsensible, are the same for everyone, and can be divided into concepts and objects.

Similar views are expressed in "Thoughts." Frege indicates there, somewhat more explicitly than in *Foundations*, that the senses alone are insufficient for knowledge of an outer world. In particular, he says,

> Sense impressions alone do not reveal the external world to us. . . . Having visual impressions is certainly necessary for seeing things, but not sufficient. What must still be added is not anything sensible. And yet this is just what opens up the external world for us; for without this non-sensible something everyone would remain shut up in his inner world. (CP 369)

This is very close to the objection to empiricism of chapter 2 of *Foundations*. If the empiricist can only take the required nonsensible something to be something psychological, then all knowledge of the external world must be based on knowledge of the inner world, and the empiricist's view collapses into subjective idealism. There is no way to escape from one's inner world. Thus, in order for Frege to argue that we can have objective knowledge, he must provide a nonpsychological account of what underlies both the grasping of thoughts and knowledge through the senses. Indeed, he says, in the next sentence from "Thoughts," "So perhaps, since the decisive factor lies in the non-sensible, something non-sensible, even without the co-operation of sense-impressions, could also lead us out of the inner world and enable us to grasp thoughts" (CP 369). I have suggested that the logical source of knowledge provides this nonsensible escape from our inner world. But what justifies the claim that the laws of logic are not themselves psychological?

Arguments against taking logic to be part of psychology appear throughout Frege's work. Most of these arguments, however, seem to presuppose both an external world and an ability to escape, in our thought, from our inner worlds. Either they proceed from the assumption that there are correct and incorrect inferences independent of the psychological laws in accord with which we judge, or they proceed from the assumption that we succeed in communicating with others. These assumptions are unobjectionable from an everyday point of view. But it is not clear that it is possible to make these assumptions and to take seriously the task of trying to show that we can have objective knowledge. If there is no escape from our inner worlds, for instance, there is no possibility of communicating with other people. The attempt to show that we can have objective knowledge cannot be taken as a central part of Frege's project. One explanation of the absence of discussions of this topic from most of Frege's writings is, as I have suggested in chapter 2, that he thought that Kant had already shown that we can have objective knowledge. However, Frege did not avoid this topic entirely.[18]

Because Frege appears, in "Thoughts," to give an explicit argument against subjective idealism, it is tempting to read this paper not only as setting out Frege's views on objectivity but also as providing his arguments that we can have objective knowledge. The problem with such a reading is that Frege simply does not say enough. A defense of the possibility of objective knowledge is surely as complicated and difficult (many would say more complicated and difficult) a project as a defense

---

[18]Although there are obvious Kantian echoes in those passages and in the view of objectivity I attribute to Frege below, I will not be talking about Kant here. My aim in this chapter, and the following chapters, is to consider the views on objectivity which can be attributed to Frege solely on the basis of his writings. As I have argued in chapter 2, it is unclear how much of Kant's work Frege studied seriously or how he read it. Thus, I am not convinced that the result of filling the gaps in Frege's explicit views with a reading of Kant will provide an epistemological perspective that can be attributed to Frege. It may well be that the apparently Kantian echoes exhaust the influence of Kant. Consequently, I have chosen to concentrate on Frege's words alone.

This is not to say that my own reading of those words can be completely independent of any other influences. The Kantian and, later, Wittgensteinian echoes that may seem to resound in these pages might well be a result of the influences writings other than Frege's have had on me. The point is that I do not want to use writings of other philosophers to support a reading of Frege's work. My aim is to give the evidence that a particular reading is accurate to Frege's writings.

of the analyticity of the truths of arithmetic. Frege devotes an entire monograph, *Foundations*, to an attempt to make it probable that the truths of arithmetic are analytic; he devotes two volumes, *Basic Laws*, to an attempt to prove this, and he devotes a number of papers to criticisms of other views about the truths of arithmetic. Yet his explicit arguments for the objectivity of knowledge appear only in "Thoughts," a short article.[19] Nor are these arguments especially original. But if Frege is not trying to provide an argument for the possibility of objective knowledge, what is he trying to do?

In order to answer this question, it is useful to remember that "Thoughts" is the first of a three-part work titled "Logical Investigations." This series of articles is yet one more attempt to communicate the nature and significance of his logic. His discussion of objectivity culminates in an argument that our knowledge through the senses must depend on something nonsensible, and this is followed by a discussion of the actuality of a thought as being constituted by the possibility of our grasping it. "Thoughts" provides another elucidation of the view that there is a substantive logical source of knowledge without which, as Frege suggests in *Foundations*, even to think at all would not be possible (FA 21). To see how this works, it may help to consider the structure of the argument from "Thoughts."

In the pages that precede those in which Frege talks about the nonsensible something that opens up the external world, he calls into question the assumptions that underlie most of his arguments against taking logic to be part of psychology. He begins "Thoughts" with a brief characterization of truth and of the laws of logic as the laws of truth. The question of truth can arise only for thoughts, that is, senses of assertoric sentences. Thus, to show that the laws of logic are not laws of psychology, he addresses the possibility that thoughts might be ideas in the subjective sense, or creatures of the mind. The discussion begins, not with a sophisticated philosophical view, but with the introduction of the views of "an unphilosophical man." These views are introduced with the following two paragraphs.

> A man who is still unaffected by philosophy first of all gets to know things he can see and touch, can in short perceive with the senses, such as trees, stones and houses, and he is convinced that someone else can equally see and touch the same tree and the same stone as he himself sees

19It was published in *Beiträge zur Philosophie des deutschen Idealismus* I, and takes up fewer than twenty pages; the English translation in *Collected Papers* takes up twenty-two pages.

and touches. Obviously a thought does not belong with these things. Now can it, nevertheless, like a tree be presented to people as identical?

Even an unphilosophical man soon finds it necessary to recognize an inner world distinct from the outer world, a world of sense-impressions, of creations of his imagination, of sensations, of feelings and moods, a world of inclinations, wishes and decisions. For brevity's sake I want to use the word 'idea' to cover all these occurrences, except decisions. (CP 360)

Why does Frege want to talk about the views of someone unaffected by philosophy?

I suspect the reason is that he thinks that as 'idea' acquires an increasingly sophisticated philosophical use, it may become less clear that to say something is an idea is to say that it is part of someone's inner world. In particular, Frege seems to think that inattention to Kant's use of this term has led to misreadings of Kant's work. In *Foundations* Frege says that the term 'idea' has both objective and subjective senses. He then says, "It is because Kant associated both meanings with the word that his doctrine assumed such a very subjective, idealist complexion, and his true view was made so difficult to discover" (FA 37). Frege's aim is to preclude the view that thoughts are ideas in the naive sense of 'idea' outlined in his characterization of the views of someone unaffected by philosophy. His argument is that, if one starts with this naive view of an inner and outer world, one will be forced to a very different view—a view that contrasts the inner world not with an outer world known through the senses but with an outer world known through the logical source of knowledge. This contrast, unlike the naive contrast, depends not at all on the assumption that we obtain knowledge through the senses. The basis of Frege's argument is the demonstration that the very notion of an idea in the subjective sense presupposes an outer world.

Frege begins by rehearsing arguments that will be familiar to readers of his earlier works. He argues that the interpretation of our talk about thoughts as talk about inner ideas will not be accurate to our understanding of these thoughts. This argument depends on the assumption that we do communicate, that our judgements are not about our ideas but about something public, that there are genuine scientific disputes. Thus, thoughts can be neither physical objects nor inner ideas. After making this familiar argument, however, Frege says,

> But I think I hear an odd objection. I have assumed several times that the same thing as I see can also be observed by other people. But what if

everything were only a dream? . . . Perhaps the realm of things is empty and I do not see any things or any men, but only have ideas of which I myself am the owner. (CP 363)

The consideration of this possibility, along with the antecedent characterization of inner ideas, leads Frege to the conclusion that he, at least, cannot be an idea and, furthermore, that it must be possible for this non-idea (himself) to be the object of his awareness. It follows that not everything is an idea and that not everything that can be the object of awareness is an idea. In this way Frege obtains an outer world without resorting to any views about the senses. It may seem misleading to use the expression 'outer world' here. After all, Frege has not obtained a physical world. Rather, he has obtained a world that is an outer world only in the sense of not being made up of his inner ideas.

This is not sufficient, of course, to license Frege's assumption that he communicates with others. He has only inferred that there is an outer world, not that the outer world has the particular character that it appears to him to have. Given this, it is possible that his everyday assumption that there are other people and that he communicates with them (hence that different people can grasp the same thought) is correct. But it does not follows that the assumption *is* correct. He says, "By the step with which I win an environment for myself, I expose myself to the risk of error" (CP 367). After all, Frege might be mistaken in his belief that there are other people and that he communicates with them. But what does this risk of error amount to?

The risk is that he might be mistaken about the character of his environment. Yet in spite of his acknowledgment that he might be wrong, Frege does not take the risk of error very seriously. He says, of the possibility that there might be other people with whom he communicates, "And, once given the possibility, the probability is very great, so great that it is in my opinion no longer distinguishable from certainty" (CP 368). How else, he asks, could there be a science of history or a distinction between natural science and fables like astrology and alchemy?

Once he has dismissed the idea that there might not be other people with whom he can communicate, the desired characterization of thoughts follows by the same sort of arguments Frege has given a number of times. Were thoughts inner ideas, there would be no possibility of communication. Throughout this argument, of course, no assumption of the existence of knowledge through senses was required. After drawing the consequence that thoughts are neither inner ideas nor objects perceptible by the senses, Frege introduces his reasons for thinking that there must be something nonsensible which opens up the external world. And given this, he adds, perhaps "something non-sensible, even without the co-operation of sense-impressions, could also lead us out of the inner world and enable us to grasp thoughts" (CP 369). The obvious candidate is the logical source of knowledge.

It may not be clear, however, that grasping thoughts involves knowl-edge. As I have argued in chapter 2, Frege views his Begriffsschrift as a means for expressing thoughts. Furthermore, the Begriffsschrift expres-sion of a thought amounts to a string of symbols which, given the rules for the use of Begriffsschrift symbols and the general logical laws, makes clear in exactly which correct inferences the thought can figure. In this way the logical source of knowledge must be involved in the grasping of a thought. This is not to say that in order for someone to understand a particular thought, she must understand Frege's Be-griffsschrift or be able to identify the general logical laws. On the other hand, she must, in some sense, understand some logical laws. Some-one, for instance, who claims that Frege is bald and that he also is not bald can be said not to understand the thought expressed by 'Frege is bald'.

Of course, Frege also says that the something nonsensible which enables us to grasp thoughts is also necessary in order for us to see instead of merely having visual impressions. But this does not tell against taking this something nonsensible to be the logical source of knowledge. Most of our everyday judgements about the physical world require general logical laws for their justification. The laws of logic, for instance, will be necessary for unwinding the specifications of the borders of the North Sea and using this information to determine whether or not the North Sea is 10,000 square miles in extent. In general, definitions of the concepts involved in an assertion about the external world along with general logical laws will be used in the justification of the assertion. But the point is not simply that our com-plex assertions cannot be justified without use of the general logical laws. Knowledge involves the judgement that a thought is true. For this the grasping of the thought, and hence the logical source of knowledge, is necessary. This may be what Frege has in mind when he says in *Foundations* (FA 99) that a logical activity is involved in observation.

The upshot seems to be that Frege is, in a sense, founding objectivity on what is required for grasping a thought. At first glance, this criterion for objectivity seems unconnected to the world in much the way Fregean meaning (*Bedeutung*), as I have interpreted it, seems unconnected to the world. After all, it may seem that what is important is not our ability to grasp thoughts but, rather, our ability to connect with the physical world—or to distinguish true thoughts about the external world from false thoughts. But Frege's point, as I read it, is that this conception is misleading. The grasp of thoughts is necessary in order to consider the question of which are true and which false. Furthermore, thoughts cannot be inner ideas. Thus, if we can account for our ability to grasp thoughts, we have accounted for our ability to connect with something other than our inner world. But connecting with something outside our inner world is not a matter of connecting with the physical world. To grasp a general thought belonging to logic, for instance, is to connect with something other than an inner idea but not with the physical world. Frege says: "Neither logic nor mathematics has the task of investigating minds and the contents of consciousness owned by individual men. Their task could perhaps be represented rather as the investigation of *the* mind, not of minds" (CP 368–369). The grasp of thoughts requires a grasp of logical regimentation—thoughts are articulated. This articulation is a feature neither of our inner ideas nor of the physical properties of the external world.

The views expressed in "Thoughts," on this reading, are closely related to those of chapter 2 of *Foundations*. In section 26, after discussing the senses in which the claims about the extent of North Sea or the color of petals of a flower are objective, Frege turns to the more general question of what is objective. He says, "I distinguish what I call objective from what is handleable or spatial or actual" (FA 35). And then he says:

> I understand objective to mean what is independent of our sensation, intuition and imagination, and of all construction of mental pictures out of memories of earlier sensations, but not what is independent of the reason,—for what are things independent of the reason? To answer that would be as much as to judge without judging, or to wash the fur without wetting it. (FA 36)

Something objective, he says, is something that is apprehended by thought, but it is not the product of a psychological process or a creature of the mind.[20]

---

[20]Hans Sluga draws the connection between these passage from *Foundations* and the above passage from "Thoughts" in *Gottlob Frege*, chapter 4, section 5. My account of

These comments fit with the comments from "Thoughts" and also with his *Foundations* comments about concepts—in particular, his denial that "concepts sprout in the individual mind like leaves on a tree" (FA vii). In fact, this talk about apprehension by thought does not seem an unreasonable criterion for the objectivity of concepts, particularly insofar as Frege insists on the scientific legitimacy of concepts under which no objects fall. But, as with Fregean meaning, our knowledge about everyday objects looks mysterious. How is this related to his claim (FA 34) that the North Sea is objective?

The North Sea is objective because, in demarcating the North Sea by drawing its boundaries (or in recognizing boundaries others have drawn), one apprehends or recognizes that object. However, this apprehension does not create the object, for, had different boundaries been drawn and had the object delimited by these boundaries been called the 'North Sea', the original object (what we in fact call the 'North Sea') would still be there, although unrecognized. In fact, the objects delimited by those other boundaries are there now, although unrecognized. It is important to realize, however, that the claim that such unrecognized objects exist need not be a substantive metaphysical claim. For, to say that something exists is simply to say that one and only one object falls under a particular concept. Drawing the boundaries of an object amounts simply to formulating a concept that picks out the object. Thus, to claim that unrecognized objects exist is simply to claim that it is possible to formulate (heretofore unformulated) concepts under which exactly one object falls. Since new concepts are formulated all the time, this is not a particularly remarkable claim. Frege gives, as examples of objective things, the earth's axis and the center of mass of the solar system. The common feature of all these objects, physical or nonphysical, recognized or unrecognized, is that their borders can be drawn, they can be distinguished from other objects.

Since it is a requirement that the drawing of boundaries for the expression 'North Sea' pick out one and only one object, the apprehension of this object is not solely conceptual. While the senses are not sufficient, they are necessary.

The drawing of borders, of course, must be expressible in words. In fact, it seems to follow from Frege's fundamental views on the relation between thought and language that our ability to recognize what is

---

Fregean objectivity bears important similarities both to his account and to the account given by Thomas Ricketts in "Objectivity and Objecthood: Frege's Metaphysics of Judgment."

objective amounts to our ability to mark it off or define it in language. To see this, it is important to recall one of the central features of Frege's formulation of Begriffsschrift, his taking judgements, as opposed to concepts, as the basic units of understanding. In one of his discussions of Boole's logical calculus, Frege describes this difference. He says, "For in Aristotle, as in Boole, the logically primitive activity is the formation of concepts by abstraction, and judgement and inference enter in through an immediate or indirect comparison of concepts via their extensions" (PW 15). Frege then discusses some of his objections to this approach before he introduces his own: "As opposed to this, I start out from judgements and their contents and not from concepts" (PW 15). His description of his view about how we get concepts is that

> we . . . arrive at a concept by splitting up the content of possible judgements. . . . We may infer from this that at least the properties and relations which are not further analysable must have their own simple designations. But it doesn't follow from this that the ideas of these properties and relations are formed apart from the object; on the contrary they arise simultaneously with the first judgement in which they are ascribed to things. (PW 17)

Even primitive, indefinable concepts, then, are arrived at through judgements in which they appear, not through some immediate, primitive recognition. And while Frege does not say here that our recognition of an object is also dependent in this way on our understanding of judgements, he makes it clear in *Foundations* that our recognition of an object and our understanding of how it is to be distinguished from other objects requires an understanding of judgements. Thus, our recognition of an object cannot be independent of conceptual knowledge or judgements. Indeed, there must be some act of cognition through which the object is recognized. And, as Frege says in passing, "every act of cognition is realized in judgements" (PW 144). For Frege, conceptual thought is dependent on language. He is quite explicit in one of his early papers, "On the Scientific Justification of a 'Conceptual Notation'," about the necessity of symbols or language for our thinking, and he says, as well,

> Also, without symbols we would scarcely lift ourselves to conceptual thinking. Thus, in applying the same symbol to different but similar things, we actually no longer symbolize the individual thing, but rather what (the similars) have in common: the concept. This concept is first gained by symbolizing it. . . . This does not exhaust the merits of symbols; but it may suffice to demonstrate their indispensability. (CN 84)

Thus, there is a sense in which the expressive powers of language underlie Frege's notion of objectivity.

It is important also to notice that Frege is trying, in this section, to distinguish what is objective from our private ideas. A (nonprimitive) object must be unambiguously picked out by some defining expression. Frege assumes that some defining expression can be given, in particular, for the earth's axis. The object defined by this expression cannot be anyone's private idea of the earth's axis. The thoughts that are communicated about the earth's axis are not claims about any particular person's idea.

There is, of course, something unsatisfying about the above argument. It seems unlikely that anyone would claim the 'the earth's axis' stands for some individual's private idea. But it may also seem that there is a more public sense of 'idea' in which we could say that 'the earth's axis' stands for an idea. While this might appear to present a problem for Frege, this public sense of 'idea' is exactly what Frege is trying to get his audience to recognize. This public or shared sense of 'idea' is simply what is communicable or can be part of our common store of thoughts. Ideas, in this public sense, are precisely what is objective.

In a footnote on page 37 of *Foundations*, Frege makes a few comments about how he understands this objective sense of 'idea' and indicates that, in this sense, numbers are ideas. Among his comments are "My arguments would be beside the point if he meant by idea an objective notion"; "An idea in the objective sense belongs to logic and is in principle non-sensible"; and "Objective ideas can be divided into objects and concepts." This objective, but nonpsychological, notion of idea is very important to Frege. He avoids using *Vorstellung* because he wants to ensure that no one will interpret his objective logical sense of 'idea' in either the subjective or the (objective) psychological sense. For the publicity of these objective ideas precludes their being objects of psychology. In Frege's objective, logical sense, the content of '2 + 2 = 4' is an idea, it is what can be common to everyone's understanding of '2 + 2 = 4', what can be communicated or shared. It is because the propositions of arithmetic have this common content that they can be objectively justified and known.

After discussing his North Sea example, Frege attempts to make this objective/subjective distinction clearer. Once more, the connection of objectivity with language is central. Everything we can think about has both an objective side (the true thoughts expressed by assertions in which it is picked out) and a subjective side (the private—and inexpressible—ideas associated with it). Frege tries to get us to distinguish between these objective and subjective realms by considering our intuition of space. Frege says, "What is objective in it is what is subject to laws, what can be conceived and judged, what is expressible in words" (FA 35). So it is an objective fact that any two points determine exactly one line; points, lines, and planes are objective, and all the axioms and theorems concerning them are objective. But our intuition of space has inexpressible features as well. In particular, we justify the axioms of geometry by the construction of mental images, and the appearance of these images is subjective. We draw pictures on blackboards in which points are represented by dots, but there is no objective relationship between the appearance of these dots and the points they represent. Frege gives an example that is designed to make the subjective nature of appearance clear. He says,

> Let us suppose two rational beings such that projective properties and relations are all they can intuite . . . and let what the one intuites as a plane appear to the other as a point, and vice versa. . . . In these circumstances they would understand one another quite well and would never realize the difference between their intuitions, since in projective geometry every proposition has its dual counterpart. (FA 35)

Frege is asking his reader to imagine the sorts of inner images each being constructs to justify statements and, in particular, axioms of projective geometry. The idea is that one can construct, for each axiom, two very different pictures, each of which is the picture that justifies the axiom for one of these beings. Now suppose these beings, each of whom has imagined one of these pictures, try to discuss their pictures. Only propositions of geometry can be stated, and the two beings agree on all of these. As long as the geometrical images that play the role of points if they appear in the mental pictures of one of the beings play the role of planes if they appear in the mental pictures of the other and vice versa, no difference between the pictures can be expressed. The inexpressible difference between the pictures, assuming that each is an inner picture of one of the two beings, is subjective. As long as they agree on the axioms and theorems of geometry, there is no objective difference.

This story may seem less odd once one observes that this story is simply a version, for geometrical knowledge, of inverted spectrum examples. These examples are designed to help us distinguish between an individual's (private) sensory experience of a color and her objective knowledge that, for instance, a flower's petals are that color. The example of the two beings who intuit projective properties and relations was designed, similarly, to allow us to distinguish the private appearance of inner intuition from the objective knowledge of geometric axioms gained from it. Thus, it is not particularly surprising that Frege goes on to argue that the sensible world has both objective and subjective sides. Claims about colors are objective, Frege says, by virtue of our ability to communicate using them, not by virtue of our ability to experience them. For even a colorblind person, someone who cannot experience the difference between red and green, can still recognize the distinction. A symptom of our conviction that color statements do not simply express the character of experience is the reasonableness of the claim that something *appears* red, but *is* white (FA 36). With this example, Frege ends his discussion of objectivity, stating that what is objective is what is independent of our sensation, intuition, and imagination, and of all construction of mental pictures out of memories of earlier sensations, but not what is independent of the reason.

CHAPTER 5

Platonism, Fregean and UnFregean

FREGE is typically associated with two views about the nature of mathematical truth: logicism and platonism. Logicism, the view that the truths of arithmetic can be proved from the laws of logic, is widely regarded today as an implausible view.[1] This is a result, in large part, of the way in which Frege's *Basic Laws* failed. In order to derive truths of arithmetic from his logical laws, Frege needed to introduce a law that enabled him to form legitimate definitions of number-names from logical symbols. But the introduction of this law, Basic Law V, made his laws inconsistent. There need not have been an inconsistency. Number-names might simply have been introduced as undefined symbols. The problem arose from Frege's attempt to formulate purely *logical* definitions of the number-names. Thus, it does not seem unreasonable to suppose that the problem with Frege's viewing numbers as logical objects does not arise from his platonism—his taking them to be objects—but from his logicism, his taking them to be definable from logical terms and truths about them to be provable from purely logical laws. And it is therefore not surprising that, unlike his logicism, Frege's

---

[1]Logicism is not universally regarded as an implausible view. David Bostock has attempted to defend a version of logicism in *Logic and Arithmetic: Natural Numbers* and *Logic and Arithmetic: Rational and Irrational Numbers*. Also, both Harold Hodes, in "Logicism and the Ontological Commitments of Arithmetic," and Crispin Wright, in *Frege's Conception of Numbers as Objects*, have attempted to defend Frege-inspired versions of logicism.

platonism is widely regarded today as a problematic but serious view.

I have identified Frege's platonism as the view that numbers are objects, but if 'platonism' is understood in a contemporary sense, this is not quite accurate. First, the term 'platonism' has been applied to a variety of distinct views, not all of which can be characterized by the identification of numbers as objects.[2] Second, in those versions of platonism which can be characterized in this way, more is required than the identification of numbers as objects—numbers must be a particular sort of object. On the characterization of platonism which seems most closely connected with Frege's texts, platonism is the view that numbers are abstract objects,[3] objects that are neither ideas nor physical objects. Much of the contemporary literature on platonism and the philosophy of mathematics is concerned with what are taken to be the consequences of this view and with attempts to resolve the difficulties they engender.[4] But are the consequences of platonism discussed in the contemporary literature truly consequences of Frege's view? There is no question that Frege claims, repeatedly, that numbers are objects but not physical objects and that the truths of arithmetic are objective truths. Insofar as these pronouncements constitute platonism (and realism), Frege was surely a platonist. But, as I have been arguing throughout this book, it cannot be taken for granted that those of Frege's pronouncements which appear to be statements of contemporary views really *are* statements of contemporary views. In this chapter I will argue

[2]See, for example, Michael Resnik's taxonomy of platonism in *Frege and the Philosophy of Mathematics*. One variety of platonism Resnik identifies, methodological platonism, requires no characterization of numbers as objects. My reason for not employing his taxonomy is that too many late twentieth-century assumptions are required if it is to be helpful. An understanding of Resnik's taxonomy requires an understanding of what it is for objects to "exist independently of us and our mental lives," for our knowledge of objects to be "based upon a direct acquaintance with them," and for the existence of numbers to be "on a par with ordinary objects." As I interpret Frege's view, these expressions have a very different sense than they have for Resnik, and when these expressions are understood in this sense, many of Resnik's distinctions collapse.

[3]It is interesting that, while many contemporary writers use the term 'abstract object' in their characterizations of platonism (see, e.g., Philip Kitcher, *The Nature of Mathematical Knowledge*, 6; Michael Resnik, *Frege and the Philosophy of Mathematic*, 162; Michael Dummett, *Frege: Philosophy of Language*, 480), Frege does not himself use such a term, except in a discussion, in section 74 of volume 2 of GGA, of Cantor's use of the term *abstracte Gedankendinge*. Frege indicates there that, in order to make sense of Cantor's use of this term, it should be understood as Frege understands *logische Gegenstände* (logical objects). In most of his writings, Frege instead talks of objects that are objective but have no place.

[4]See, for instance, Penelope Maddy, "Perception and Mathematical Intuition"; Mark Steiner, *Mathematical Knowledge*.

that the consequences of Fregean platonism, of taking numbers to be nonsubjective, nonphysical objects in Frege's sense, are very different from those discussed in the contemporary literature. In particular, I will argue that Frege is not concerned with realism or platonism in any contemporary sense and that, from his point of view, contemporary debates about realism, at least with respect to arithmetic, make very little sense.[5]

Although Frege does not use any expression that can be (or has been) translated by 'abstract object' (except in his discussion of Cantor's use of *abstracte Gedankendinge*, GGA vol. 2, 86, section 74), it is not difficult to identify the statements from *Foundations* which look to be Frege's assertions of platonism. In section 61 (FA 72), Frege claims that numbers are objects that are neither physical objects nor ideas.[6] Frege also appears to advocate a similar sort of platonism for certain nonmathematical objects. For, in his discussion of the senses of assertoric sentences (thoughts) in "Thoughts," he describes thoughts as neither things in the external world nor ideas (CP 363). He then goes on to claim that there is a "third realm" in addition to that of the external world and the realm of subjective ideas. Frege argues that neither numbers nor thoughts are created by us and our thinking. It is tempting to view Frege's assertions as answers to the following question: If numbers (thoughts) are things, then which things are they? And Frege's answer—that numbers (thoughts) are nonphysical objects that exist independently of us—seems to require an antecedent metaphysical picture on which the world is made up not only of physical objects, but also of nonphysical objects. This account of the background against which Frege's claims about numbers in section 61 of *Foundations* should be read looks to be further supported by the question to which he turns after making these claims. Section 62 begins with the question "How, then, are numbers to be given to us, if we cannot have any ideas or intuitions of them?" (FA 73). The appearance of this question suggests that there are special epistemological problems about our access to or knowledge about nonphysical, nonsubjective objects—problems that do not arise if our concern in restricted to physical objects. What, exactly, are these problems?

[5]Some other arguments that Frege's concern with realism is not a concern with realism in a contemporary sense appear in Cora Diamond, "What Does a Concept-Script Do?" and Hans Sluga, *Gottlob Frege*, 106.

[6]See also FA 58, 69–70.

In a paper titled "Mathematical Truth," Paul Benacerraf gives a contemporary articulation of the epistemological problems involved in the metaphysical picture of an objective world containing both physical and nonphysical objects. Benacerraf says that what is needed, in order to account for our knowledge of arithmetic, is "an account of the link between our cognitive faculties and the objects known."[7] The reason this is needed to account for our knowledge of arithmetic but not for our knowledge of the physical world is that we already know what the requisite link between our cognitive faculties and physical objects is. Physical objects are given to us through the senses. Thus, if numbers are abstract objects, it seems important to ask, with Frege, how numbers are given to us.

Kurt Gödel, the writer who, besides Frege, is most often associated with a contemporary mathematical platonism, has written something that looks to be a direct response to this worry. In "What Is Cantor's Continuum Problem?" Gödel says that

despite their remoteness from sense experience, we do have something like a perception also of the objects of set theory, as is seen from the fact that the axioms force themselves upon us as being true. I don't see any reason why we should have less confidence in this kind of perception, i.e., in mathematical intuition, than in sense perception.[8]

That is, to the question "What provides the link between us and the numbers?" Gödel answers "mathematical intuition." Benacerraf suggests both that there is something right about Gödel's attempt at a solution and that there is something that makes it unsatisfactory. What is right about Gödel's response to the puzzle is the postulation of "a special faculty through which we 'interact' with these objects" (MT 416). But Benacerraf worries that Gödel's analogy between this mathematical intuition and sense perception is too superficial, for we still need an account of how the axioms force themselves upon us as being true.

Is Frege led, by similar considerations, to ask the question that begins section 62? If so, there is something very peculiar about Frege's answer. Unlike Gödel, Frege does not explicitly identify a special faculty that provides the requisite link between us and mathematical objects.

---

[7] Paul Benacerraf, "Mathematical Truth" (MT), 415. All further references to this work in this chapter will be made parenthetically in the text.
[8] Kurt Gödel, "What Is Cantor's Continuum Problem?" 483.

Frege's immediate response to the question is to say, "Since it is only in the context of a proposition that words have any meaning, our problem becomes this: To define the sense of a proposition in which a number word occurs" (FA 73).[9] But will such definitions provide an account of a faculty by which we interact with the numbers? Certainly in *Foundations*, Frege does not explicitly identify a faculty that provides such a link. And it is not clear how the sort of definitions Frege wants could be used for this identification. His definitions of the senses of sentences containing number words contain words only for logical notions. Moreover, the object-expressions that appear in Frege's restatements are all of the form "the extension of the concept . . . ," and Frege says nothing about interaction with, or a link to, extensions. Indeed, all he does say is "I assume that it is known what the extension of a concept is" (FA 80). If numbers are abstract objects, on Frege's view, then extensions of concepts must be abstract objects. So if Frege needs an account of a faculty by which we interact with abstract objects in order to answer his question, his definitions, as characterized in *Foundations*, fail to provide such an answer.

Why should Frege have asked the question that begins section 62 if he did not mean to provide a satisfying answer? The explanation, I will argue, is that, insofar as platonism is taken to be the kind of view which gives rise to Benacerraf's puzzle, Frege is not a platonist.[10] Benacerraf's puzzle is not a contemporary articulation of the question that begins section 62. To see this, it will help to examine not only the passages in which Frege claims that there are objects that are neither physical objects nor subjective ideas but also the passages that appear to be responses to something like Benacerraf's puzzle. Passages of the latter sort will provide clues for the interpretation of Frege's claims that there are objective nonphysical objects—that is, clues for an understanding of what Frege's 'platonism' comes to.

It may seem puzzling that, though Frege raises the question about how the numbers are given to us in *Foundations*, he does not give an answer there that identifies a faculty that gives us the numbers; he does appear to give an answer in the second volume of *Basic Laws*, however,

---

[9]In many of his writings, Frege uses the term 'proposition' (*Satz*) ambiguously. Sometimes he uses it to talk about verbal or written expressions of thoughts (sentences), and sometimes he uses it to talk about thoughts. (See Frege's comment, COR 182.) In this particular passage, he seems to be talking about sentences.

[10]Here I am not alone. Sluga, in *Gottlob Frege*, and Gregroy Currie, in *Frege: An Introduction to His Philosophy*, have given interpretations on which Frege's project is primarily epistemological and platonism plays no role in his overall project.

where he says, in a discussion of logical objects, among which are the numbers, that we grasp them through our logical faculties (*logische Fähigkeiten*, GGA vol. 2 section 74). This appears to be addressed to the sort of question Benacerraf asks. The senses, after all, are the source of our knowledge of truths about the physical world, and Frege is intending to show, in *Basic Laws*, that the logical source of knowledge is the source of our knowledge of the truths of arithmetic. Since, as I have argued earlier, a notion of logical faculties (the logical source of knowledge) plays a role in the epistemological picture that motivates the project of *Foundations* and *Basic Laws*, the analogy between the senses and the logical faculties seems less ad hoc than Gödel's analogy between the senses and mathematical intuition.

Furthermore, Frege's answer to the question about how the numbers are given to us—that we grasp logical objects through our logical faculties—may seem to avoid the superficiality Benacerraf attributes to Gödel's answer. For at least a part of an answer to the question of how the primitive logical laws and, by inference, the truths of arithmetic force themselves upon us as being true is available to Frege. Frege suggests in *Foundations* that if we try to deny a fundamental truth of arithmetic, complete confusion ensues and it may no longer even be possible to think at all (FA 21). On its own, this claim may seem hyperbolic. In the context of Frege's account of the content of number statements in *Foundations*, however, it looks more understandable.

Frege's account provides a sketch of how everyday truths of arithmetic can be proved from primitive logical laws.[11] This involves restating

---

[11]Although Basic Law V does not appear in *Begriffsschrift*, the *Foundations* sketch makes use of an assumption that is tantamount to Basic Law V. The assumption, in more contemporary language, is that the holding of an equivalence relation between two objects can also be construed as an identity. That is, the claim that two objects stand in an equivalence relation is also a claim that there is some*thing* which is the same. A similar point is made in the second volume of *Basic Laws*. Frege argues, in section 147 of the second volume, that mathematicians have, in fact, made use of this assumption. Mathematicians, Frege notes, will replace the claim that, for any argument f and g have the same values, or:

$$(x)(fx = gx)$$

with the claim that f and g are identical, or:

$$f = g.$$

The latter expression is, on Frege's view, improper. Functions cannot be said to be identical or not identical. Frege says, "Accordingly, the symbolism '$f = g$' cannot be recognized as correct; but nevertheless it shows that mathematicians have already made use of the possibility of our transformation" (TWF 160).

the fundamental truths of arithmetic in language containing only logical terms. Given the restatements of these truths, their status as logical truths is meant to be evident. Furthermore, the denial of a fundamental truth of arithmetic requires more than the bald claim that it is false. In order seriously to deny the truth of some fundamental arithmetical claim, one must be prepared to accept the logical consequences of this denial. The denial of primitive logical laws will be among these consequences. And, as I have argued earlier, Frege thinks the truth of the primitive logical laws is forced on us by thought itself. Thus, any attempt to consider the consequences of the denial of a truth of arithmetic will, unlike apparently similar attempts to consider the consequences of the denial of an axiom of geometry, involve us in contradicting ourselves. The obvious absurdity into which we would be forced by a serious denial of a fundamental truth of arithmetic will, in this sense, make further thought impossible. Perhaps even more significant, thinking involves grasping thoughts and grasping a thought involves an understanding of its logical form, that is, of the inferences in which it can figure. But such an understanding is not possible if some of the primitive logical laws are denied.

Can we then view Frege's logical faculty as a special faculty by which we interact with abstract (or logical) objects? There is, at least, something odd about calling this a "special" faculty. For the logical faculty is not simply introduced as a means for interaction with abstract objects, it is involved in *all* thought and inference. Frege makes it clear repeatedly that the logical faculty is involved, not simply in a limited field, but in all our objective knowledge.

Is it important that the faculty that allows us to interact with abstract objects be a "special" faculty? In order to answer this question, it may help to look more closely at Benacerraf's puzzle. This puzzle is supposed to derive its force from an account of differences between our access to physical and abstract objects. We are asked to notice that we have no recognizable access to abstract objects, *unlike* physical objects—to which we have access through the senses. Why is this difference significant? The asssumption is that our claim to objective knowledge must be defended, at least in part, by an account of our interaction with or access to objects. The senses, given this requirement, provide a piece of an account of our knowledge of the physical world, while the corresponding piece is missing for an account of our knowledge of the world of arithmetic. The significance of this difference, then, is tied to a view about how our claim to objective knowledge must be defended. But who holds this view?

Benacerraf does not claim to be describing a problem that affects everyone who might be inclined to say numbers are abstract objects. He means, rather, to be describing a problem that affects only those who want to combine this claim with some sort of causal theory of knowledge—a theory that, he says, "follows closely the lines that have been proposed by empiricists" (MT 413).[12] This is interesting in light of Frege's almost immediate dismissal of empiricist answers to his questions. But it does not, of course, follow that Benacerraf's problem is not a problem for Frege. I think, however, that the assumptions that make Benacerraf's problem worrying for him are assumptions that Frege would not endorse. To see this, let us examine Benacerraf's characterization of his own epistemological views.

Benacerraf characterizes the views of knowledge in which he is interested as any views on which "Hermione can learn that the cat is on the mat by looking at a real cat on a real mat" (MT 412). After introducing the notion of a causal account of knowledge, Benacerraf claims, in support of this sort of account, that it seems to be the right way to account for our knowledge about medium-sized objects. He adds, "Furthermore, such knowledge (of houses, trees, truffles, dogs and bread boxes) presents the clearest case and the easiest to deal with" (MT 413). He goes on to say that other cases of knowledge can be explained by tying them via inference to these clear cases.

If it is possible to give an account of knowledge along these lines, however, the faculty that enables us to interact with abstract objects really must be special. For Benacerraf's assumption is that we can begin an account of knowledge with an account of our knowledge about medium-sized objects. That is, Benacerraf assumes that an account of our knowledge about medium-sized objects does not depend on an antecedent account of our knowledge about abstract objects. The completion of an account of our knowledge of medium-sized objects will enable us to see what is required of an account of our knowledge of other sorts of things. He assumes, further, that we already have enough

---

[12]The connection between empiricism and suspicion of abstract objects is made explicitly by Wright in *Frege's Conception of Numbers as Objects*. Wright also notes that the argument for platonism in *Foundations* leaves a great deal of work for the reader, and Wright proceeds to do some of this work. The work required, for Wright's purposes, is an investigation of the extent to which the argument for platonism in *Foundations* can be reconciled with empiricism. Wright goes on to make a number of illuminating points both about empiricism and about a platonism that is consistent with Frege's apparently platonist remarks. Although I am in agreement with many of these points, my interest is in fitting Frege's remarks not into an empiricist framework but into the framework of his own epistemological picture.

of such an account to see what some of the requirements will be. For knowledge about abstract objects we will need a special faculty that bears the sort of relation to these objects which the senses bear to physical objects.

It should now be clear that Frege rejects one of the fundamental assumptions on which the reasonableness of Benacerraf's epistemological strategy depends. It is not merely that, on Frege's view, our knowledge of medium-sized objects does not present the clearest cases of knowledge. Frege does not believe that the senses alone can give us any objective knowledge of the physical world.[13] He says, "We need the perceptions, but to make use of them, we also need the other sources of knowledge" (PW 268). What is also needed is language or a way of drawing boundaries. Frege's view is that the physical world is not articulated—that we impose structure on it, and our knowledge of physical objects is mediated by this structure. It is not that, without imposing articulation via language, we could not interact with physical objects, but rather that, without imposing this sort of articulation, our interaction with physical objects cannot play a role in our knowledge about the external world. Sense perceptions may provide us with links to physical objects, but sense perception alone does not provide us with intelligible experience.

There is no reason to suppose that Frege would deny that Hermione can learn that the cat is on the mat by looking at a real cat on a real mat. But this looking, this observation, involves a logical activity as well. It involves the grasping of a thought and it involves the employment of logical laws. In order for Hermione to judge that the cat is on the mat, she must divide up the world into cats and non-cats, mats and non-mats. As I argued in chapter 4, this dividing up is not, on Frege's view, something given to us via the senses by the physical world.[14] Hermione needs, in fact, to impose structure on the world.

The same "something non-sensible," Frege suggests, is involved both in the escape from our inner world we achieve by seeing things (as opposed to having mere visual impressions) and in the escape we achieve by grasping thoughts (CP 369). If the logical source of knowledge is needed for judgements about abstract objects, it is also needed for judgements about physical objects. Thus, we cannot, on Frege's

[13]Sluga makes this point in *Gottlob Frege*, 105.
[14]I take Dummett to be making much the same point when he says, in *Frege: Philosophy of Language*, that, for Frege, "the world does not come to us articulated in any way; it is we who, by the use of our language (or by grasping the thoughts expressed in that language), impose a structure on it" (504).

view, provide an account of our knowledge of medium-sized objects without an account of the logical source of knowledge. The logical source of knowledge cannot be the sort of special faculty Benacerraf requires.

The senses truly do constitute a special faculty. They constitute what is needed, in addition to the logical faculty, that is, in addition to thought, in order for us to have knowledge about physical objects. But the logical faculty, far from being a special faculty, is needed for all objective judgement. What would a special faculty, which plays a role in our knowledge of abstract objects analogous to that played by the senses in our knowledge of physical objects, be? It could only be something that needs to be added to the logical faculty in order to give us knowledge of abstract objects. But this knowledge requires nothing in addition to the logical faculty. On Frege's view our knowledge about numbers, *unlike* our knowledge of physical objects, requires no special faculty.

My account of why Frege has not answered Benacerraf's puzzle may seem, at this point, to focus too closely on the role of a special faculty and not closely enough on the notion of access to or interaction with objects. Benacerraf's requirement that there be a special faculty for our interaction with abstract objects does seem to be a consequence of the empiricism in his epistemological picture. He starts with a view on which the world is presented to us already divided into objects. Thus, for him, no logical faculty is necessary for dividing up the world—we need only account for our access to those objects and the divisions that are already out there. Part of this account, the part that is needed for everyday medium-sized objects, is immediately available. What he wants is an analogous account for other objects.

On Frege's view, however, there is no immediately available account for our knowledge of everyday medium-sized objects and, consequently, no reason to take an account of knowledge about this sort of object as a paradigm for an account of knowledge in general. But if there is no reason for Frege to take this sort of account as a paradigm, this does not show that Frege's views about knowledge do not require an account of our interaction with or access to the objects about which we can have knowledge. Is there some point in Frege's account of the source of our knowledge about the numbers at which he needs to account for our interaction with numbers? In order to answer this, it will help to review the outlines of Frege's account of the source of our knowledge about the numbers.

The questions that, on Frege's picture, count as epistemological have to do solely with the justification of judgements. Earlier I argued that Frege's answer to the question beginning section 62, "How . . . are numbers . . . given to us?", is very odd if this question is understood as the sort of question Benacerraf is asking. But if the considerations that give rise to Frege's question concern only justification, his answer does not seem odd at all. Frege's answer amounts to an explanation of what sort of primitive truths will have to figure in any justification of thoughts that are expressed by sentences in which number-words appear. It is that the only sort of primitive truths needed are primitive logical truths. If the logical faculty or logical source of knowledge is sufficient to give us the primitive logical truths, then it will be sufficient to give us truths about numbers. This answer is given more explicitly in volume 2 of *Basic Laws*, where Frege writes:

> If there are logical objects at all—and the objects of arithmetic are such objects—then there must also be a means of apprehending, of recognizing, them. This service is performed for us by the fundamental law of logic that permits the transformation of an equality holding generally into an equation. (TWF 161)

If the means of apprehending logical objects are provided by a fundamental law of logic, then it seems that the faculty by which we have access to the logical objects is the logical faculty or the logical source of knowledge—the faculty by which we recognize the truth of the fundamental law of logic. Does Frege assume that recognition of a fundamental law of logic enables one to interact with objects?

The sleight of hand, if there is any, will have to do with the justification of the primitive logical law that gives us objects. Up to this point, I have made no distinction between the status of objects given to us by logical laws in the pre–*Basic Laws* writings and their status in *Basic Laws* and the post–*Basic Laws* writings. In fact, there is no evidence that Frege's underlying epistemological picture changed over this period. However, over the period between the publication of *Foundations* and the publication of the first volume of *Basic Laws*, an important change takes place in Frege's logic.

In Frege's first exposition of his logic, in *Begriffsschrift*, there is no logical law that gives us objects. In fact, from Frege's point of view both at the time of publication of *Begriffsschrift* and at the time of publication of *Foundations*, there are no admissible proper names that appear in *Begriffsschrift*. The letters used in the proofs and definitions given in *Begriffsschrift*—whether the role they play in the expression of inferences is that of object-expressions, function-expressions, or expressions for contents that can become judgements—are all used to express generality.[15] Everything Frege proves in *Begriffsschrift* is completely general. Of course, from Frege's later point of view, there are proper names in the proofs that appear in *Begriffsschrift*. For among the changes that occur between the publication of *Begriffsschrift* and Frege's later writings are the introduction of the sense/Bedeutung distinction and the classification of sentences as proper names. In *Basic Laws* there is also a further change. The notation introduced in *Basic Laws* has symbols for a second-level function that converts first-level functions (including concepts) into objects. The introduction of these symbols allows Frege to formulate a new primitive law of logic, Basic Law V, which permits, as Frege says in the quotation above, "the transformation of an equality holding generally into an equation" (TWF 161). Thus, it looks as if the question of how the logical source of knowledge can give us objects must be addressed differently for the period before the new logical law of *Basic Laws* was formulated and the period after which this logical law was formulated.

If, at the time of publication of *Foundations*, Frege did not have a primitive logical law that enabled him to apprehend objects, why was he confident that the logical source of knowledge would be sufficient to give us the numbers? It seems that there is only one way to answer this. It seems that Frege must come up with a concept for each number. That is, for each number, he must find a concept for which he can show that anything falling under that concept has the features that the number in question must have and for which he can show that it holds of precisely one thing. Under these circumstances, a definite description locution will enable him to come up with a definition for each numeral. Frege responds to the question of section 62, of how numbers are to be given to us, by saying that what is needed is an explanation of how we can derive a criterion, for each number, which will determine, for any object, whether or not that object is the number in question. Such a criterion, for each number, will determine the requisite concepts. But where will these criteria come from?

In section 63 he explains that his strategy will be to take the notion of identity as already known and to use this "as a means for arriving at that which is to be regarded as being identical" (FA 74). In order to give his readers a sense of this strategy, he turns, in section 64, to a discussion of parallel lines and direction. He begins as follows,

[15]See the discussion in section 1 of *Begriffsschrift*.

The judgement "line *a* is parallel to line *b*", or, using symbols,

   a // b

can be taken as an identity. If we do this, we obtain the concept of direction, and say: "the direction of line *a* is identical with the direction of line *b*". Thus we replace the symbol // by the more generic symbol =, through removing what is specific in the content of the former and dividing it between *a* and *b*. We carve up the content in a way different from the original way, and this yields us a new concept. (FA 74–75)

The crucial assumption is that the judgement that two lines are parallel can also be taken as an identity. In effect Frege's assumption appears to be that any judgement that can be expressed as the holding of an equivalence relation can also be expressed as an identity.[16] Frege makes a similar assumption for the judgement that there is a one-one correlation between the Fs and the Gs. In the case of the parallel lines, the identity is between the direction of line *a* and the direction of line *b*; in the case of the concepts, the identity is between the number of Fs and the number of Gs. But, Frege observes, it is not sufficient to say that what is the same is the direction of (number of) something unless we guarantee that there is precisely one direction of (number of) and are clear about what this one thing is.

Unless we guarantee that there is precisely one object that is the number of something, there are no grounds for ruling out the claim that Julius Caesar and the moon are both numbers of some concept. We also have to make sure that the description of what the number of a concept is is sufficiently explicit to determine whether or not Julius Caesar is the number of the concept. To make sure that this is guaranteed, Frege introduces the notion of extension and then gives the following, as a definition: "The Number which belongs to the concept F is the exten-

[16]Actually, Frege talks not about equivalence relations but, rather, about transforming an equality holding generally into an identity (see, e.g., BLA 44, 72; TWF 159–160). I have chosen to talk about equivalence relations instead, because it is not clear that when Frege talks about the existence of a one-one correlation between concepts in *Foundations* he understands this to be an equality holding generally. Of course, the claim that there is a one-one correlation between the Fs and the Gs is not, as stated, obviously a claim that an equivalence relation holds. Frege actually restates this by introducing a relation for which he introduces the term *gleichzahlig*, which Austin translates as 'equal'. Frege then restates the claim that the Fs can be correlated one-one with the Gs as: the concept *F* is equal to the concept *G*. The expressions 'the concept *F*' and 'the concept *G*' are object-expressions, and the relation that is said to hold between them is an equivalence relation. I will address the issue of Frege's view of 'the concept *F*' and 'the concept *G*' as object-expressions and, hence, as designating objects, in chapter 6.

sion of the concept 'equal to the concept F' " (FA 79–80). He goes on to show that this definition works. That is, he goes on to show that the extensions have the features the numbers must have. He does not, however, go on to show that the expression "extension of the concept 'equal to the concept F' " holds of precisely one object, or to explain why none of these extensions is identical to Julius Caesar. For this requires no demonstration. He only says, in a footnote, "I assume that it is known what the extension of a concept is" (FA 80).

The assumptions in *Foundations* about our knowledge of extensions and about the possibility of transforming certain judgements into identities are informal versions of Basic Law V. These assumptions suffice to commit Frege to the inconsistency that cannot be proved from Frege's logical laws in *Begriffsschrift* but can be proved from the logical laws of *Basic Laws*. It is also clear from Frege's discussion in this part of *Foundations* that he did not feel the need to introduce any special faculty that allows us to interact with logical objects (in this case, extensions). Had Frege held the sort of epistemological views which underlie the Benacerraf picture, he would have been unable simply to assume that it is known what the extension of a concept is. As it is, he makes this assumption because he takes the introduction of extensions as merely providing a term for expressing what is going on in a basic mathematical move—the move from a generalization of an identity to a particular identity. This move, in fact, appears in his own mathematical work. In his doctoral dissertation, "On a Geometrical Representation of Imaginary Forms in the Plane," he says,

> Taken literally, a 'point at infinity' is even a contradiction in terms; for the point itself would be the end point of a distance which had no end. The expression is therefore an improper one, and it designates the fact that parallel lines behave projectively like straight lines passing through the same point. 'Point at infinity' is therefore only another expression for what is common to all parallels, which is what we commonly call 'direction'. (CP 1)

In *Basic Laws*, he claims that mathematicians make comparable moves and, insofar as they do so, implicitly recognize extensions (see GGA vol. 2, sections 146–148; TWF 159–161). The general acceptability of this move in mathematics, on Frege's view, licenses his claim that it should be regarded as a primitive logical law as well as the consequence that our knowledge of extensions comes from the logical source of knowledge alone.

There is no sleight of hand in Frege's account of our knowledge of the truths of arithmetic because, on his view, the justification of the primitive logical laws on which our apprehending mathematical objects depends does not require a relation between us and mathematical objects. Frege's logic is meant to capture the content of an assertoric sentence which is relevant to legitimate inference, and one basic type of legitimate inference, on Frege's view, is the inference from certain sorts of generalization to identities.

I have argued that Benacerraf's characterization of the epistemological problems involved in viewing numbers as abstract objects breaks down when it is applied to Frege's epistemology. But I have not argued that there is no sense in which Frege's claim that numbers are objects can get him into difficulties. Furthermore, if Benacerraf's characterization of the kind of interaction needed between us and numbers does not describe anything to which Frege's epistemological views commit him, it does not follow that Frege's view requires no interaction of any kind. In particular, I have ignored one very notable way in which we are widely regarded to interact with numbers—we refer to them. How do we succeed in talking about the numbers?

If Frege does need an account of how we refer to the numbers, Benacerraf may be on the right track. That is, there may be a special problem for abstract objects which is not a problem for physical objects. In the above discussion I have argued that our knowledge of physical objects is no easier to explain, on Frege's view, than our knowledge of nonsubjective, non-spatio-temporal objects. It may seem that I now need a similar argument for reference. However, I will not proceed to give such an argument. It is not that I think that, for Frege, our reference to physical objects is any easier to explain than our reference to non-spatio-temporal, nonsubjective objects. Rather, it is that, as I read Frege, the possibility of reference, unlike the possibility of knowledge, is not something that requires explanation.

This should not be especially surprising. I have already argued that Frege's notion of a term's having Bedeutung amounts to the demarcation of sharp boundaries for the application of the term and the indication of the epistemological roots of truths expressed by sentences in which the term appears. I have also argued that, to show that a term has Bedeutung, it suffices to show that it is either primitive or definable in the appropriate way from primitive terms. On this characterization, Frege's Bedeutung is precluded from playing the role of reference, that is, from playing a role in an explanation of how words or other linguistic entities hook onto extralinguistic entities.

Frege's repeated assertions that an object-expression or proper name has Bedeutung just in case it stands for an object look to create a serious tension for the above characterization. His conditions for an object-expression's having Bedeutung seem simultaneously too strict and too loose to be the conditions under which an object-expression stands for an object. The strictness of Frege's conditions is not obviously a problem. I have argued that Frege's criteria for a proper name's standing for an object are sufficiently strict that they are not met by some of our everyday proper names which, on our everyday understanding of what it is for a name to stand for an object, clearly do stand for objects. On the other hand, Frege is claiming that our everyday language is insufficiently precise for scientific contexts, and he might well want to suggest that we ought to modify our everyday views about the circumstances under which a proper name stands for an object. The more serious problem is that it seems that satisfaction of Frege's conditions, the requisite sharpness of demarcation of the expressions used in a science, may not be strict enough to guarantee that scientific language avoids wandering into fiction.

Although Frege's criteria for definitions are designed to guarantee that an appropriately defined proper name stands for something provided all the expressions from which it is defined do so, there is no guarantee that a primitive proper name stands for something. All that is required for the introduction of a primitive term is elucidation designed to eliminate vagueness and ambiguity—elucidation that might, in any case, go wrong. It might seem, then, that via the introduction of primitive proper names that have precisely understood sense but that, in fact, do not stand for any objects, a portion of some science could turn out to be fictional.

This consequence looks to be a reductio of Frege's claim that a proper name is admissible for scientific purposes just in case it stands for an object. It is important to note, however, that this consequence depends on the assumption that Frege would have allowed proper names to be among the primitive terms of some systematic science. But it is by no means obvious that such an assumption is warranted.

It is important, first of all, to realize that if Frege prohibits the introduction of primitive proper names, this does not preclude the employment of proper names in scientific contexts. Proper names need not be formed by definitions from primitive proper names. Although a number of proper names are introduced in *Basic Laws*, for instance, no primitive proper names appear. The only primitive symbols of the *Basic Laws* Begriffsschrift are expressions for first-, second-, and third-level functions.[17] Frege's most famous proper names are formed by the use of the second-level function that takes first-level functions as arguments and whose value, for a first-level function, is the course of values of that function. Other notable proper names are formed using a scientifically admissible version of a definite description function.[18] Furthermore, the use of a concept-expression that has Bedeutung, in Frege's sense, cannot lead us inadvertently into fiction. A concept-expression has Bedeutung just in case it determinately holds or does not hold of each object. Concepts that either hold of everything or hold of nothing are perfectly appropriate for scientific purposes because they have sharp boundaries. The mere use of a concept-expression cannot lead us into fiction. We may mistakenly claim that some thing falls under the concept. But, if so, the mistake is that the claim is false, not that it is fictional.

Thus, the possibility of our satisfying Frege's requirements on the admissibility of a term for science yet ending up with some proper name that does not stand for any object is dependent on either

(1) the permissibility of the introduction of primitive proper names or
(2) the possibility of giving an acceptable definition of a proper name using admissible concept-expressions which fails—that is, the possibility that the proper name so defined does not stand for any object.

The second possibility is not really a problem. In the discussion below, it will become apparent how Frege's requirements on definitions prevent the sort of situation described by (2) from occurring. My main concern in the following pages will be with the issue of the permissibility of primitive proper names.

If it is possible to satisfy Frege's requirements for the admissibility of a term for science yet end up with some proper name that does not stand for an object, then there will be a role for a reference relation to play. A sentence containing a proper name that does not stand for an object will express a fictional thought; it will be neither true nor false. The aim of

[17]It may seem that Frege's *Basic Laws* relies on a hidden primitive proper name, 'the True'. But it is important to note that to say, as Frege does, that the notion of truth or of being true is primitive is not to say that the True is a primitive logical object. Tyler Burge makes this distinction in "Frege on Truth." For an argument that 'the True' is not a hidden primitive proper name, see Appendix A.

[18]This is a function whose value is stipulated for cases in which the concept whose extension is involved does not hold of precisely one object. I say here "the concept whose extension is involved" rather than "the concept involved" because the definite description function is a first-level function, not a second-level function.

Frege's requirements on the admissibility of a term for purposes of science is to preclude the possibility of formulating fictional assertions as well as vague or ambiguous assertions in a language for a systematic science. The worry suggested above is that Frege's requirements may not be sufficient to guarantee that his proper names stand for objects. If this is so, what needs to be added to Frege's requirements is a requirement that his primitive proper names refer to objects. To satisfy this further requirement, of course, will be to show that we can have a particular sort of interaction with these objects—that we can refer to them. Thus, if Frege would permit the introduction of primitive proper names into the language of a systematic science, he would be committed to showing that we can have this sort of interaction with them. If he would permit the introduction of primitive proper names into arithmetic, he would be committed to showing that we can interact with numbers.

Would Frege permit the introduction of primitive proper names into the language for a systematic science? Frege does not give an explicit answer to this question. He does, however, provide us with one language for a systematic science. This language is his Begriffsschrift. In the first version of Begriffsschrift there are no proper names at all. In the second version, the version introduced in the first volume of *Basic Laws*, there are proper names but no primitive proper names. All proper names of Begriffsschrift are constructed out of the names of primitive first-, second-, or third-level functions. Of course, it by no means follows, from the fact that there are no primitive proper names in Begriffsschrift, that the introduction of primitive proper names is prohibited for any construction of a systematic science, or even that primitive proper names are prohibited in logic.

How significant, then, is the absence of primitive proper names from Frege's logical notation? A primitive logical proper name would have to be a name for a logically simple object, and truths about this simple object would have to be provable from logical laws alone. Given Frege's views about logic, there are only two candidates for primitive logical objects: the True, the object that is named by all true sentences, and the False, the object that is named by all false sentences. Frege says, in one of his discussions of logical simplicity,

> If something has been discovered that is simple, or at least must count as simple for the time being, we shall have to coin a term for it, since language will not originally contain an expression that exactly answers. On the introduction of a name for something logically simple, a definition

is not possible; there is nothing for it but to lead the reader or hearer by means of hints to understand the word as is intended. (CP 182–183)

This appears to fit with his remarks about truth. He concludes a discussion of truth in "Thoughts" with this remark: "So it seems likely that the content of the word 'true' is *sui generis* and indefinable" (CP 353). The claim that the notion of truth is primitive is repeated in a number of unpublished papers and fragments.[19] And truth is clearly a logical notion; the laws of logic are the laws of truth. Could Frege introduce a primitive proper name for the True into his logical notation?

Frege's comments about the primitiveness of truth, as well as his lack of an explicit description of the True and the False, suggest that we need primitive names for the True and the False. But not only does Frege include no primitive name for the True in his notation, he actually says, in *Basic Laws*, "It is always possible to stipulate that an arbitrary course-of-values is to be the True and another the False" (BLA 48). He then proceeds to provide such stipulations. How is this to be reconciled with his comments about the indefinability of the notion of truth?

It is important to notice that when Frege talks about the indefinability of truth, he is talking about what is meant by the predicate 'is true' rather than the True. Furthermore, Frege provides no evidence that what is meant by 'is true' is the True or that the simplicity of the former shows that the True is logically simple. In general, Frege views concepts as having associated objects. But the fact that a concept is simple does not entail the simplicity of the associated object. Something logically simple cannot be defined. And, by means of the second-level extension function, it *is* possible to define the extension of a simple concept. I do not mean to be suggesting that the True is the extension of the concept 'is true', but only that it does not follow from the simplicity of the predication of truth that what Frege wants to call 'the True' is a logically simple object. It may also be helpful to think of a truth-value (i.e., the True or the False) as what is common to two thoughts whose biconditional is true—just as a direction is what is common to parallel lines, a truth-value is what is common to two sentences that express thoughts whose biconditional is true. Frege's claim that an arbitrary course-of-values can be stipulated to be the True suggests that the notion of what is common to two true thoughts, like the notion of what is common to two parallel lines, is insufficiently precise for scientific purposes. I will

[19]See, e.g., PW 126, 129, 174.

discuss Frege's stipulation of courses-of-values for the True and the False later.[20] The important point for now is that since, on Frege's view, it is not possible to define something logically simple, and since Frege explicitly defines the True and the False, he is not taking the True and the False to be logically simple objects.

Assuming Frege's logical notation is adequate for expressing the logical laws, there cannot be any primitive logical proper names. Indeed, since the only other a priori science is geometry, none of whose primitive signs is a proper name, there cannot be primitive proper names in any a priori science. This does not seem entirely unreasonable. Logic, after all, is the most general science. No logical truth depends on any assertions about particular objects (FA 4). Nor do the truths of geometry, whose systematic construction contains no primitive proper names, depend on any assertions about particular objects. In empirical sciences, on the other hand, there looks to be more of a role for proper names to play. Particular facts are used as inductive evidence for general laws.

This evidence leaves open two possibilities. Either Frege would not have permitted primitive proper names in the language of *any* systematic science, or he would not have permitted primitive proper names in the language of any a priori science but would have permitted them in the languages of some empirical sciences. Instead of attempting to determine which of these was Frege's position, I will consider how each position will influence the answer to the question with which I began this discussion: Does Frege need an explanation of how we succeed in talking about numbers?

I will begin by supposing Frege would not permit any primitive proper names in any language for a systematic science. If so, an object-expression will stand for an object only if it is appropriately constructed from precise concept-expressions. An appropriate construction, of course, must guarantee that the expression stand for some object. If I were to introduce a name (say, 'Gottlob') via a definite description (say, 'the *F*'), I would be required to stipulate an object for which the name stands in the cases in which either more than one object is an *F* or no object is an *F*. I might stipulate, for instance, that in such cases the expression will stand for o.[21] The appropriate sort of definition would

---

[20]See Appendix A.

[21]Frege suggests this stipulation for divergent series in "On Sense and Meaning," CP 169. Of course, the strategy for definite descriptions suggested here will work only if there

guarantee that 'Gottlob' stood for some object. Thus, if 'Gottlob' has been appropriately introduced, it *will* stand for some object. The answer to the question "How do we succeed in referring to Gottlob?" is simply that 'Gottlob' has been appropriately defined. It appears, then, that there is no question to ask about the relation between an object and a linguistic expression. But if there is no question to be raised about whether or not 'Gottlob' stands for an object, there are still important questions to be raised.

A likely reason for introducing 'Gottlob' via the description 'the *F*' is that we suspect that there is exactly one *F* and that there is some interest in investigating the one and only *F*. So there are important questions to ask. These important questions concern what sort of thing Gottlob is. Is Gottlob an *F*? And if Gottlob is an *F*, how many *F*s are there? For, if Gottlob is not the one and only *F*, there is a sense in which, although our expression does stand for an object, it has failed to do what we wanted it to do. Thus, the real success of our definition of 'Gottlob' depends on whether or not there is one and only one *F*. But this success has nothing to do with language or reference. This success has to do with the concepts and assertions involved in the systematic science in whose language 'Gottlob' was defined. The questions about how many *F*s there are are no different from the sort of question a researcher with no special interest in language might ask. Consider: Is there a planet in our solar system beyond Neptune? Or: How many planets are there in the solar system beyond Neptune? The upshot is that there are no special questions to be asked about how our object-expressions hook onto the world, provided they have been appropriately defined.

Let us turn to the numerals and the question: How do we succeed in talking about numbers? Suppose the numerals and the expression 'number' are appropriately defined. Since the numerals are object-expressions, there will be no question that they stand for objects. But the numerals might stand for the wrong sort of objects. We need to be

---

is at least one proper name that is not defined via definite description. In *Basic Laws*, Frege's definite description operator is a first-level function. That is, it takes objects as arguments. Frege's actual definition stipulates that the value of the definite description function is identical with the argument except in the case that the argument is an extension with precisely one member. In this case, the value is the member of the extension. To use Frege's definite description operator in forming a name that will do the work of 'the *F*', one applies Frege's function to the extension of *F*. If there is one and only one *F*, then the value of Frege's function, applied to the extension of *F*, is the one and only *F*. In any other case, the value of the function is the extension of *F*.

sure that the numerals really stand for numbers. The justification of our claims to success in talking about the numbers, however, will consist of nothing more or less than a series of proofs. We need to show that what we now mean by '3', for instance, really does satisfy the characteristics that numbers must satisfy. For example, we will need to be able to prove, using the definitions of '3', '1', and '>' that 3 is greater than 1. And we will need to use the appropriate definitions to show that 3 is the number that belongs to the concept *is greater than or equal to 2 and less than or equal to 4*. We will also need to use the definition of 'number' to prove such general claims as the associativity of addition, and so on. In other words, what we will need, in order to show that the numerals stand for the numbers, is precisely what Frege intends to show in *Basic Laws*. We will need to use Frege's definitions to prove the basic truths of arithmetic.

That Frege lacks an account of a reference relation seems, on the view described above, entirely reasonable. For there is no question that can be sensibly asked and that has anything to do with a reference relation. It is not just that the idea of giving some metaphysical account of extralinguistic objects and an account of how the numerals hook onto these objects is entirely foreign to Frege's enterprise. The significant point is that, if no primitive proper names are permitted in any systematic science, then, once all expressions of the language of a systematic science are introduced in a manner Frege would consider appropriate, the only substantive scientific questions will be just like the questions considered above. It will be pointless to ask whether a particular object-expression stands for something because, by virtue of having been appropriately introduced, it does stand for something. On the other hand, we can ask to what it refers. But this question is not a question about the existence of an object "out there" to which the expression bears a reference relation. It is, rather, a question about what sort of thing the object that the expression stands for is. The answer to the question will involve the same sort of everyday research in which researchers with no interest in language, semantics, or a reference relation routinely engage.

Consider, for instance, a definition of the term 'Pluto' as the planet in our solar system beyond Neptune. In order to meet Frege's constraints in the formulation of such a definition, it will be necessary to stipulate what 'Pluto' stands for if there is no planet beyond Neptune or if there are more than one. Given this stipulation (and assuming all the terms that are constituents of the definition to be either primitive or appropriately defined), there is no reasonable question to be raised about how we succeed in referring to Pluto or about whether the term 'Pluto' has a reference. We can, of course, ask "To what does 'Pluto' refer?" But this is not a question about language or semantics. It does not call for investigation into the definition or the nature of our relations to the external world. Rather, it calls for investigation of our solar system. The questions that must be answered are: Are there planets beyond Neptune? How many planets beyond Neptune are there? There is no role to be played in this investigation by a general account of circumstances in which proper names refer to objects. For similar reasons, there is no role to be played in Frege's investigation by such an account. Just as it will be important for the Fregean definer of 'Pluto' to ask whether there is a planet in our solar system beyond Neptune, Frege might ask whether there is a smallest prime number greater than 1,000. The question of whether "the smallest prime greater than 1,000" refers is easy to answer, but empty. The question of whether the smallest prime greater than 1,000 is a prime greater than 1,000 which is smaller than every other prime greater than 1,000 amounts to the question of whether or not there is a prime that is greater than 1,000 and smaller than all other primes greater than 1,000. This question requires, for its answer, no metaphysical theory or theory of reference, but only a mathematical existence proof. The question of whether there are abstract mathematical objects will be settled likewise. Insofar as the issue of platonism or realism for mathematics concerns whether or not there really are objects "out there" for which our number-names stand, there will not be a substantive issue for Frege.

There is one point in the above argument which might seem worrying. The only complex proper names I have discussed are definite descriptions. But if the view described is to work, it may seem that there must be at least one proper name that is not a definite description. For, in order to guarantee that all definite descriptions stand for objects, it is necessary that each definite description carry with it a stipulation of an object for which it is to stand if the concept used in the definite description either holds of no object or holds of more than one object. If each definite description requires such a stipulation and if each object must be picked out by a definite description, it seems that the attempt to define a proper name will involve an infinite regress.[22] Frege is not

[22]Although Frege is not committed to this consequence, since not all his proper names are definite descriptions, there may seem to be a way out for someone who recognized

committed to this consequence, however. For empirical sciences, the object used in the stipulative clause can simply be a logical object, say o. Thus, the question, for Frege, can be reduced to the question of how he guarantees that his logical proper names stand for objects. These proper names include not only definite descriptions, but also proper names formulated from combinations of first- and second-level function-expressions. The guarantee that these combinations stand for objects is a consequence of Basic Law V and Frege's stipulations in section 10 of *Basic Laws*.

Of course, Basic Law V is inconsistent with the other logical laws and the argument of section 10 of *Basic Laws* is notoriously problematic.[23] These are significant, perhaps insuperable, barriers to defending the adequacy of Frege's requirements on the admissibility of a term for science for accomplishing his aims. However, my purpose here is not to defend these requirements. Rather, my purpose is to determine whether or not there is a role to be played by a reference relation within Frege's picture. The existence of significant problems with the views that allow Frege to talk about a proper name's standing for an object without introducing the notion of a reference relation does not show that Frege had a notion of reference. Nor does Frege's ultimate repudiation of Basic Law V show this.

The above argument is based on the assumption that Frege would not have allowed any primitive proper names in any language for a systematic science. But, as I mentioned earlier, there is no conclusive evidence that Frege would not have allowed primitive proper names in the

---

only definite descriptions as proper names. One might introduce the primitive concept-expressions, demonstrate that a particular concept holds of precisely one object, and use the expression for this concept to formulate a definite description without stipulating a value for the case in which the concept does not pick out precisely one object. This definite description would be the only one permitted which does not have such a stipulation, and this definite description would be used for the stipulative clause in all other definite descriptions. In this case, the nonfictional status of the science would depend on the demonstration that the concept used for the stipulative clauses holds of precisely one object. It is not clear that such a solution would be acceptable to Frege, however. In particular, if there is any flaw at all in the research that licensed the use of the original definite description, then this language would, in Frege's words, "have the fault of containing expressions which fail to designate an object (although their grammatical form seems to qualify them for that purpose) because the truth of some sentence is a prerequisite" (CP 168–169). Frege indicates that this situation "arises from an imperfection of language" (CP 169).

[23]See, for instance, A. W. Moore, and Andrew Rein, "*Grundgesetze*, Section 10"; Charles Parsons, "Frege's Theory of Number."

empirical sciences. Let us now turn to the second hypothesis: that Frege would have allowed primitive proper names, but only in the empirical sciences. If this is so, and if Frege wanted his requirements for a term's admissibility for science to apply to all science, then he would have needed to introduce a reference relation and to use an account of this relation to supplement his requirements for a term's admissibility for empirical sciences. That is, there would have to have been room, in Frege's picture, for an account of how proper names hook onto objects that are independent of us. If there is room for such an account, how would this affect Frege's views about numerals and the requirements necessary for our success in referring to numbers?

The role for a notion of reference in Frege's picture, if it exists, is a consequence of the difference between empirical and a priori science. Primitive proper names can be introduced only if a definition cannot be given. If we need to use primitive proper names to talk about the empirical world, this marks an important difference between physical objects and nonphysical, nonsubjective objects. The difference is that, in an important sense, physical objects are *less* accessible to us than nonphysical, nonsubjective objects—at some point our concepts fail to pick out particular physical objects and we are forced to rely on primitive proper names. But there is no comparable difficulty for nonphysical, nonsubjective objects—our mathematical existence proofs work as before. It is only the difficulty of talking about physical objects which allows the question of how our words can hook onto physical objects to be raised seriously.

There is no reason to think that the introduction of a reference relation would force Frege to give up all his earlier views about Bedeutung. Presumably, the account of the reference relation would fit in with the nonproblematic cases of a term's having Bedeutung and the reference relation would be used only for supplementing the conditions for admissibility in the problematic cases, the cases of primitive proper names. Thus, the requirements for admissibility for concept- or function-expressions and complex proper names in which no primitive proper names occur would be as before. If so, there would still be no substantive way to raise questions about how we succeed in talking about the numbers other than Frege's original way. For he makes use of no primitive terms. That is, his success in talking about numbers can be demonstrated only by providing the requisite definitions—those Frege calls for in *Foundations*—and the proofs Frege intended to give in *Basic Laws*.

A suspicion that something is missing from my account may remain. For the above discussion ties the truths of arithmetic to the concepts we grasp, and one might suspect that the view that emerges is an antirealist view. On the other hand, Frege repeatedly makes remarks that look to be straightforward defenses of realism. I will now turn to an examination of some of these remarks. My aim will not be to classify Frege's view as realist or antirealist but only to show how the apparently intralinguistic requirements on the admissibility of a term for science fit remarks that, had they been written by a philosopher in the late twentieth century, might be read as advocating some sort of realism for numbers.

Frege repeatedly makes comments in *Foundations* which seem to indicate that truths about numbers are eternally true and are true independent of us and our thought (see FA vi–vii, 60, 90–91). We do not, on Frege's view, create the numbers: "The mathematician cannot create things at will, any more than the geographer can; he too can only discover what is there and give it a name" (FA 108). In later writings Frege is even more explicit. He says, for instance, about the statement "3 is a prime number,"

> What we want to assert in using that proposition is something that always was and always will be objectively true, quite independently of our waking or sleeping, life or death, and irrespective of whether there were or will be other beings who recognize or fail to recognize this truth. (CP 134)

And he says:

> Numbers do not undergo change, for the theorems of arithmetic embody eternal truths. We can say, therefore, that these objects are outside time; and from this it follows that they are not subjective percepts or ideas. (CP 230)

There are several ways in which these passages appear either to conflict with the views I have attributed to Frege or at least to require more of Frege's notion of Bedeutung than I have claimed is required.

First, the claim that we can only discover what is there and give it a name suggests that the world has a structure independent of any structure we impose on it. Indeed, it might be taken to suggest that this structure of the world determines our concepts. But if my interpretive difficulty with the first passage can be cleared away, this may create a difficulty for reading the second passage. For, if, as I have argued, on Frege's view the world is articulated by the structure we impose on it, it is not clear how there can be truths that are independent of our waking or sleeping, life or death, and so on. Finally, if the world is independent of us, the intralinguistic criteria for having Bedeutung I have described seem insufficient for guaranteeing that we succeed in talking about the world. Although the passages quoted above all have to do with truths about numbers, each of these issues has to do, not with the special nature of abstract objects, but with our ability to succeed in talking about a world that is independent of us. I will consider these issues in order.

The first has to do with the interpretation of Frege's claim that both geographer and mathematician can only discover what is there and give it a name. This may seem to suggest that the world (both physical and abstract) already is divided up into objects, that we interact with those objects (at least insofar as we discover them), and that this interaction is required for giving the objects a name. But that suggestion is dependent on the assumption that the discovering we do is independent of any structure imposed on the world. Yet surely the geographer is a perfect example of someone who cannot discover what is there and give it a name without relying on an antecedent structure imposed on the world. Not only does the geographer's discovery presuppose a description of what counts as a mountain, the discovery presupposes a literal map. For the geographer's discovery is not just the existence of a mountain; it is also the existence of a mountain in a certain location. The geographer's task requires some antecedent part of a map of the world, some antecedent system for locating the objects she or he can discover and name. Without such a map there would be no discovery and, in particular, no way of recognizing the mountain as the same mountain on another occasion.

In his correspondence with Jourdain, Frege gives an example of geographical knowledge (COR 80). This is the discovery that a mountain called 'Aphla' by one group of people is the same mountain as that called 'Ateb' by another group of people. The discovery is made possible by the determination of the mountain's position on maps by two different explorers and the later comparison of these maps.

Frege's analogy between the geographer's and the mathematician's tasks does not have the force of the claim that the numbers are non-spatio-temporal objects that are, in some sense, "out there" to be discovered, independent of any structure our concepts impose on the world. The "there", in the description of the geographer's task as discovering what is there, is dependent on her antecedent map of the world. That is, it is dependent on a structure people have imposed on the world for their purposes. The details of the map itself, that is, the coordinate system within which places and objects are located, are not determined by the world. There is no reason to suppose that the same people could not have chosen a different way of mapping the world. But this no more shows that location on the map is a subjective idea than the observation that the boundaries of the North Sea might have been drawn in different places shows that the North Sea is a subjective idea. Once this map is imposed, the issue of what there is in a particular location is not up to the geographer to decide; she can only discover what is there.

Similarly, discovering what is "there" in the "realm of the abstract" amounts to discovering what meets the descriptions that interest us. We don't reach out with some ineffable hook to a realm of the abstract and discover that the hook has attached itself to, for instance, infinitely many primes. Rather, having formulated the notion of a prime number, we can ask—and answer via proof—the question of whether there are infinitely many primes. It is true, of course, that the formulation of the concept 'prime number' is, on Frege's view, an act of discovery. That is, in formulating this concept, we do not construct some mental image that is then the subject of our discussions of primeness. Rather, our formulation allows us to talk about an objective concept, a concept that is the same for everyone.

There is also, of course, an important disanalogy between the task of the geographer and the task of the mathematician. The discovery of what is in a particular location on a map of the physical world requires not simply a map but also empirical evidence. It will be necessary to explore the location in question in order to determine what is there. But any talk about exploring the 'realm of the abstract' is metaphorical—the realm of the abstract has no location. I have argued that the notion of discovery, for this 'realm', amounts to the formulation of definitions and proofs. Thus, Frege says, of Cantor's symbol ∞: "In terms of our definitions this has a perfectly clear and unambiguous sense; and that is enough to justify the use of the symbol ∞ and to assure it of a meaning" (FA 96–97). And he says: "But it is also true that we have no need at all to appeal to any such sense-perception in proving our theorems. Any name or symbol that has been introduced in a logically unexceptionable manner can be used in our enquiries without hesitation" (FA 97). But it is not clear that the metaphor will bear any more weight than this. To say that abstract objects are just like physical objects, except that they are not physical, is uninformative unless the respects in which abstract objects *are* like physical objects are spelled out.

If, as I have argued, Frege believes the world is not articulated but we impose structure on it, how can he also say there can be truths that are independent of us? The answer is that, although the articulation is imposed by us—the world does not dictate where the borders of the North Sea are drawn, whether anyone draws any borders for any seas, or even whether the concept of being a sea is formulated—this articulation is objective. That is, given a particular formulation of the concept of being a sea, particular objects either are or are not seas—no mental images or thought processes can make the state of Wisconsin into a sea.

This is not to say that, prior to someone's formulating the concept of being a sea, the state of Wisconsin might have been a sea. The formulation, by some person, of the concept of being a sea is necessary only for the entertainment, by some person, of the thought. That is, the formulation (or apprehension) of the concept is a part of what is necessary for someone to grasp the thought that Wisconsin is not a sea and, in order to judge that Wisconsin is not a sea, this thought must be grasped. But this is not any part of the ultimate grounds on which the truth of the thought rests. No challenge to the claim that Wisconsin is not a sea can be met by an account of someone's apprehension of the concept of being a sea. The justification of the claim that the state of Wisconsin is not a sea does not depend in any way on any account of when the concept of being a sea was formulated nor does it depend on the existence of people who formulated this concept. Rather, the justification depends on the content of that concept, on what it is to be a sea.

Thus, although our way of regarding the world does play a role in our formulation of the claims and, hence, in our actual judgement of their truth, it does not play any role in their justification and, hence, is not part of their actual thought content. Because of this, Frege can say "That a statement of number should express something factual independent of our way of regarding things can surprise only those who think a concept is something subjective like an idea" (FA 60).

I have not entertained such counterfactual situations as those in which no humans ever existed or those in which every human dies tomorrow, because I do not know how to spell out the appropriate details of such stories. But Frege himself does not engage in much storytelling, and most of what he does say is easily explained without recourse to storytelling. As noted, he says that 3 is prime is true "independently of our waking or sleeping, life or death, and irrespective of whether there were or will be other beings who recognize or fail to recognize this truth" (CP 134). Although one might attempt to explain this by spelling out a number of science-fiction scenarios, such effort seems unnecessary if Frege's concern is taken to be with justification. For the justification of the claim that 3 is prime requires a gapless proof from primitive truths. Such a proof (Frege assumes) will contain no claims about our waking or sleeping, about anyone's life or death, about the past, present, or continuing existence of sentient beings. He also says, in *Foundations*, that one of the conditions necessary to arrive at his definition of the number 1

> is, for example, that blood of the right quality must circulate in the brain in sufficient volume—at least so far as we know; but the truth of our last proposition does not depend on this; it still holds, even if the circulation stops; and even if all rational beings were to take to hibernating and fall asleep simultaneously, our proposition would not be, say, cancelled for the duration, but would remain quite unaffected. (FA 91)

Once again, this can be explained by noting that the proofs of the a priori propositions require no claims about rational beings.

I now turn to the last of the three interpretive difficulties introduced above. If the world is independent of us, then it seems that the satisfaction of the intralinguistic criteria for having Bedeutung I have described cannot suffice for Frege to account for our success in talking about the world. The intralinguistic criteria require that to have Bedeutung a term must either be defined in an appropriate way from primitive terms or be primitive itself. If the primitive terms stand for things in the world, then Frege's criteria for definitions will guarantee that all terms defined from primitive terms also stand for things in the world. Thus, if the primitive terms stand for things in the world, we *do* succeed in talking about the world. But, as I mentioned earlier, there is no guarantee that primitive terms stand for things in the world. Frege says, about establishing the meaning of primitive terms:

> The purpose of explications is a pragmatic one; and once it is achieved, we must be satisfied with them. And here we must be able to count on a little goodwill and cooperative understanding, even guessing; for frequently we cannot do without a figurative mode of expression. But for all that, we can demand from the originator of an explication that he himself know for certain what he means; that he remain in agreement with himself; and

that he be ready to complete and emend his explication whenever, given even the best of intentions, the possibility of a misunderstanding arises.

Since mutual cooperation in a science is impossible without mutual understanding of the investigators, we must have confidence that such an understanding can be reached through explication, although theoretically the contrary is not excluded. (CP 301)

One notable feature of this description is that what seems to be at issue is not a connection between the primitive terms and a world that is independent of us but, instead, agreement among users of the primitive terms. For Frege, however, this comes to the same thing. If we have a common understanding of a primitive term, then we can use it to ask questions about the world. If primitive terms are not object-names, there is no question about whether or not we succeed in talking about the world. The only question is whether our claims about the world are true or false.

None of this is meant to show that there is no relation between Frege's Bedeutung and contemporary notions of reference. Indeed, if a proper name has Bedeutung in Frege's sense, it will have a referent and its Bedeutung will be its referent. Michael Dummett says, "It is precisely because the expressions we use have such extra-linguistic correlates that we succeed in talking about the real world and in saying things about it which are true or false in virtue of how things are in that world" (FPL 198). I do not think Frege would disagree. My point, however, is that on Frege's view—at least for the statements of arithmetic, and perhaps for all statements—there is no way to make sense of the possibility that we do not succeed in talking about the real world. This is not to say that we cannot fail in our descriptions of the world. But this failure is not due to our descriptions not being descriptions of the world but to their inaccuracy as descriptions of the world. The possibility of failure is the possibility that those statements we have judged to be true are either false or insufficiently precise. Once we see that Frege's notion of Bedeutung is not meant to play a role in providing an account of how words hook onto the world, we can also see that Frege's intralinguistic characterization of Bedeutung leaves nothing out.

I began this chapter with an argument that if Frege is a platonist, he is not the sort of platonist described by Benacerraf, nor would he be bothered by the problems sketched by Benacerraf in "Mathematical Truth." The argument was based on an account of Frege's anti-empiricism, his view that the world is not pre-articulated, that we impose an articulation on the world. In the last few pages I have addressed the apparent tensions between this view and some of the passages from Frege's writings which look to be expressions of realism. It is now time to address another sort of passage which has been used to bolster the attribution of contemporary platonism to Frege, the sort of passage in which he claims there are nonsubjective, nonphysical objects. If such a claim itself constitutes platonism, then Frege surely is a platonist. But I have already argued that certain views that are traditionally associated with platonism cannot be attributed to Frege. I now turn to the issue of what Frege's claim that there are nonsubjective, nonphysical objects comes to.

Although, in my discussion of Fregean Bedeutung, I examined the apparent conflation of the precise definability of an object-expression and its standing for an object, I wrote as if there were no issue of how to understand what Frege means when he talks about a term's standing for an object. But if I have given the impression that there really is no issue—that Frege's only departure from everyday language has to do with his use of Bedeutung and its cognates—it is important to correct this impression. In order to understand what Frege can mean when he writes about a term's standing for an object, it is necessary to understand what he means by 'object' (Gegenstand). Since some of Frege's comments about a term's standing for an object are clearly meant to exploit our everyday notion of the name/bearer relation, it is tempting to regard Frege's use of 'object' as a generalization of the everyday notion of medium-sized physical objects. And it is only a short step from such a view to the conviction that Frege needs an answer to the sort of question Benacerraf asks. But Frege's claim that numerals stand for objects need not be understood as a claim that numbers are just like physical objects except that they have no location.

It is important to note, first of all, that when Frege actually discusses his use of 'object', he does not try to generalize on the notion of everyday medium-sized objects. For instance, after noting the oddness of his designation of truth-values as objects in "On Sense and Meaning," he does not proceed to draw analogies between truth-values and everyday medium-sized objects but instead says "What I am calling an object can be more exactly discussed only in connection with concept and relation" (CP 163–164). Frege's intertwined notions of objecthood and concepthood are discussed in greater detail in "On Concept and Object," where he says that, although there are a variety of uses for the word 'concept', his use is a logical use. The logical use of both 'concept' and 'object' is clearly what he has in mind in "On Sense and Meaning" when he says, "Places, instants, stretches of time, *logically considered*, are objects, hence the linguistic designation of a definite place, a definite instant, or a stretch of time is to be regarded as a proper name" (CP 170, my emphasis). The content of Frege's claims that something is an object (or concept) can be cashed out via an explication of the logical regimentation of what can be said about it in a logically perfect language.

There is, of course, an obstacle to defending this characterization of Frege's claim that something is an object. On Frege's view, the notions of object, concept, and relation are simple and cannot be defined. Thus, the significance of Frege's claim that numbers are objects will depend on his elucidatory statements about how 'object' is to be understood. But Frege's hints about how to understand the words 'object', 'concept', and 'relation' are notoriously problematic, and a detailed consideration of this issue now would probably have the effect of obscuring the structure of the argument about Frege's platonism. For these reasons, I have saved the issue of whether or not Frege's notion of objecthood is coherent for chapter 6. For now I will give a brief sketch of how Frege understands 'object' and try to show how this notion is connected with Frege's claims that numbers are objects.

Frege's notions of objecthood and concepthood are simple logical notions. Our recognition of the legitimacy of his logical notation and the inferences it licenses depend on our antecedently sharing a common store of thoughts. And, in order to share a common store of thoughts, we must share a common view of correct inference. For our understanding of a thought is largely an understanding of how it can figure in inferences. For instance, someone who would deny the legitimacy of inferring that 1 + 1 is a prime number from the claims that 2 is a prime number and that 2 is 1 + 1 could not be said to understand the thoughts involved. In order to recognize the legitimacy of this inference we must be able to divide the thoughts involved into their logical constituents. In the inference described above, the thought expressed by '2 is a prime number' can be divided into the sense of the proper name '2' and of the concept-expression 'is a prime number'. Furthermore, we must recognize the difference in the logical role played by different constituents. Frege expresses this difference metaphorically by saying that the sense of 'is a prime number', the concept-expression, is unsaturated, while that of '2', the proper name (or object-expression), is saturated.[24] Un-

---

[24]See, for instance, PW 191–193, CP 280–281.

saturated senses pick out concepts, while saturated senses pick out objects.

Frege also says, of the difference between objects and concepts, "One could express it metaphorically like this: There are different logical places; in some only objects can stand and not concepts, in others only concepts and not objects" (CP 282). This metaphor can be partially cashed out by noting that the result of replacing the occurrence of '2' in some sentence with 'is a prime number' will be gibberish. On the other hand, the result of replacing the occurrence of '2' with '3' may be false, but it will not be gibberish. Frege's requirements for a logically perfect language may be viewed as a generalization of this. He requires that, for any sentence that expresses a thought, that has a truth-value, and in which a proper name appears, the result of replacing that proper name with any other (admissible) proper name expresses a thought that has a truth-value (although perhaps a different truth-value from the original thought). In other words, what can be predicated of an object can be predicated of *any* object.

The metaphor can also be viewed as having been cashed out in the rules of Begriffsschrift. The rules of Begriffsschrift tell us which combinations of Begriffsschrift symbols and letters are permissible. For instance, in section 19 of the first volume of *Basic Laws*, Frege explains which letters can be used to express something holding generally of functions (i.e., function-variables) and which can be used to express something holding generally of objects (object-variables). He also says:

> In general after a function-letter there follows within its scope a *bracket* whose interior contains either one place, or two separated by a comma, according as the letter is to indicate a function of one argument or of two. Such a place serves to receive a simple or complex sign that denotes or indicates an argument, or occupies the argument-place in the manner of the small Greek vowels. (BLA 70)

The unsaturatedness of the function-letter is indicated by the brackets, which must be filled in in any series of Begriffsschrift symbols expressing a complete thought.

The above characterizations of Frege's views about objecthood and concepthood make no use of any analogy between everyday medium-sized objects and nonphysical objects. Although there is no explicit account of what it is to be an object (or concept), no such account is necessary for us to be able to recognize objects as objects. If we know anything at all about some object, then this knowledge must be expressible in words. And our understanding of what these words express requires an understanding of the logical articulation of what they express. What is common to everyday medium-sized objects and non-physical objects is simply that they share a logical place. What can be said about one sort of object in a logically perfect language can be said about (although perhaps not be true of) the other. It should be clear that the recognition of this feature that all objects share does not depend on considerations concerning what sort of entity everyday physical objects are. Although everyday physical objects are certainly examples of objects, Frege's notion of object can be derived by considering features of these objects only if the features considered are logical features—that is, the way in which senses that pick out these objects figure in the articulation of thoughts and the way in which thoughts about these objects figure in inferences. And in order to do this the thoughts must be expressed in language.

If this exhausts the significance of Frege's notion of objecthood, his claim that numbers are objects can be understood as a claim about the role that must be played in inferences by the senses of number-names in a systematic science that can replace arithmetic. But to draw this conclusion from Frege's general accounts of objecthood alone would be precipitous. It is important also to examine how this account of objecthood fits the arguments that numbers are objects in *Foundations*. Frege argues, in chapter 3 of *Foundations*, that the content of an assertion of number (i.e., an answer to the question "How many . . . ?") is an assertion about a concept. And it is this claim about the content of assertions of number which provides the constraints on his definition of the concept of number. He begins work on this definition in chapter 4, which is divided into four parts, the first of which is headed "Every individual number is a self-subsistent object."

The argument that every number is a self-subsistent object begins not with an analogy between numbers and physical objects but with strategies for defining the numbers 0 and 1. In the first three sections, 55–57, Frege examines the roles played by 0 and 1 in assertions. The sentences in which number-words occur are of two types, those in which number-words play the grammatical role of proper names, and those in which they play the grammatical role of adjectives. Adjectives are, generally, concept-words. This, for Frege, illustrates the inadequacy of grammar as a guide to logical regimentation. On Frege's view, proper names and concept-words require different sorts of definitions; a word can be defined only once, and any definition of a number-word must be usable in the explanation of the content of any sentence in which it appears. Thus, to succeed, he must either provide restatements of sentences in which number-words appear adjectivally or provide restatements of sentences in which they appear as object-names.

Frege gives two considerations in favor of restating the content of sentences in which number-words appear as adjectives. One of these is that, unlike the adjectives that really are concept-expressions, numerals never appear as the entire predicate. For instance, brownness can be predicated of an object on its own, or it can constitute only a part of what is predicated. We can say both 'Frege had a brown horse' and 'Frege's horse is brown'. On the other hand, although 'two' can appear as a part of a predicate as in, for instance, 'Frege had two horses', twoness cannot be predicated on its own. To see this, consider Frege's earlier argument that assertions of number are assertions about concepts. If we view twoness as a property, then in the above example, twoness applies not to Frege's horse but to the concept *Frege's horse*. But if we try to express this by finding a sentence in which 'two' appears on its own we will end up either with 'The number of Frege's horses is two' or 'The number two belongs to the concept *Frege's horse*'. In the former case, we do not have a simple predication but, rather, an identity. For we can infer from this, along with the identity $2 = 1 + 1$, that the number of Frege's horses is $1 + 1$. In the latter case, 'two' is only a part of the predicate.

The other consideration in favor of restating the content of sentences in which number-words appear adjectivally is that his concern is "to arrive at a concept of number usable for purposes of science" (FA 69). In statements of pure arithmetic numbers play the logical role of objects. Numerals appear flanking the identity sign, and from $1 + 1 = 2$, along with something that can be truly predicated of 2, we want to be able to infer that the same thing can be predicated of $1 + 1$.

This completes Frege's direct argument that numbers are self-subsistent objects. In effect, the argument is that the definitions of the number-words which will satisfy his criteria of adequacy—that is, which will allow us to provide a systematic science that can replace arithmetic (and hence a Fregean legitimation of the pre-Fregean inferences of arithmetic)—will be definitions of proper names. There is nothing in these sections to indicate that the claim that numbers are self-subsistent objects has any more content than this.

In the next four sections, 58–61, Frege answers criticisms of his claim that numbers are objects. He considers two objections. The first is that numbers cannot be objects because we have no mental images of them, and the second is that numbers cannot be objects because they have no location. Although Frege agrees both that we cannot have mental images of the numbers and that numbers have no location, he argues that neither of these facts shows that numbers are not objects. In both sections 59 and 60, the two sections in which he discusses the significance of our not having mental images of the numbers, he talks about inferences. He argues that our failure to have any mental images of numbers does not prevent us from making inferences—the fact, for instance, that we have no mental images of our distance from the sun does not prevent us from calculating this distance and using this calculation in further inferences. He also argues that our mental images *should not* (and do not) play a role in inferences. In many cases not only are our mental images inaccurate but we are aware of this inaccuracy. Our image of the Earth as a moderate-sized ball does not prevent us from making judgements about its (nonmoderate) size.

This emphasis on inference and judgement is exactly what one would expect, given Frege's later elucidations of objecthood and concepthood. In fact, the way in which he writes about our knowledge of the earth in these sections suggests that the fact that the earth is an object is not to be gleaned from the character of our unarticulated experiences or mental images of it. Rather, the objecthood of the earth is evident from what we say and infer about it; from our saying, for instance, that the earth is a planet. We need linguistic articulation and concepts in order for our experiences to allow us to make judgements about the earth. Thus, Frege reiterates his context principle: "But we ought always to keep before our eyes a complete proposition. Only in a proposition have the words really a meaning" (FA 71). And he goes on to say,

> The self-subsistence which I am claiming for number is not to be taken to mean that a number word signifies something when removed from the context of a proposition, but only to preclude the use of such words as predicates or attributes, which appreciably alters their meaning. (FA 72)

Our license to claim that the earth is an object is a result both of our legitimate inferences about the earth and of the role the expression 'the earth' plays in the expression of these inferences. Similarly, our license to claim that numbers are objects is a result of the role number-words play in the expression of legitimate inferences.

On this view there is no reason to suppose that each object must have a spatial location. What is objective is what is expressible in words, that is, what is communicable or intersubjective. What is an objective object is what occupies a certain sort of logical place in inferences. Thus, in response to the challenge that numbers have no spatial coordinates, Frege simply replies, "Not every object has a place" (FA 72). This view of what it is to be an objective object also explains why Frege replies as he does to the next question he raises: How are numbers to be given to us? An object is given to us through the justification of our judgements and our understanding of them. In some cases, if the object is an everyday medium-sized object, for instance, our senses may be required both in order to grasp thoughts about it and in order to support the judgements we make that certain thoughts are true. But in all cases the logical source of knowledge is required, for in all cases we must understand sentences and how they can function in the expression of legitimate inferences. To recognize something as an object involves dividing up some thought into its constituents. In the case of numbers, Frege means to show, our senses are not required. What is required is an understanding of the content of scientific claims in which number-words appear.

Frege argues that, in order to replace the inferences of arithmetic and its applications with inferences of a systematic science, we must have number-names that function as proper names. Thus, insofar as platonism requires a characterization of abstract objects based on an analogy with physical objects, Frege's argument that numbers are objects is not a platonist argument. It is also important to notice that there is nothing in these sections which indicates that Frege thinks numbers are any less objective, any less independent of our minds, than physical objects. For when Frege says that the self-subsistence he claims for number doesn't mean that a number-word signifies something when removed from the context of a proposition, he is *not* thereby distinguishing number-words from other proper names. This claim appears in his discussion of the significance of our mental images of physical objects; in this discussion, only two paragraphs earlier, he makes this claim not about numerals, but about words in general.[25]

[25]Resnik says, about this passage and the passage from chapter 2 of *Foundations* where Frege says that we cannot say what things are independent of the reason (FA 36), "All of this leads the reader to believe that Frege's numbers are not full-fledged entities, that the objectivity of arithmetic—like that of geometry—does not result from the existence of a mind-independent domain of numbers which can be apprehended by many, and the numerals cannot be assigned their references through baptism or ostension or any process analogous to that through which, say, proper names of persons acquire reference" (*Frege and the Philosophy of Mathematics*, 164). Since Frege's comments in both of the

Thus far I have concentrated on Frege's *Foundations* assertions about the objecthood of numbers. It has been suggested, however, that Frege's platonism, in a contemporary sense, did not emerge until later.[26] The evidence for this is largely that many of Frege's later writings contain claims about a domain or realm of nonphysical objects existing independently of our recognizing them. Among these independent objects are numbers and thoughts. It will be useful to consider the case of numbers and thoughts separately.

As I argued earlier, there is nothing in Frege's view that numbers are given to us through the logical source of knowledge which would force him to deny their independence of us. For their independence of us amounts to nothing more nor less than that the justification of truths of arithmetic (or other truths involving numbers) requires no facts about us or our ideas. Frege says in the introduction to *Basic Laws*, "In the same way, that which we grasp with the mind also exists independently of this activity, independently of the ideas and their alterations that are a part of this grasping or accompany it" (BLA 23). To say that the existence of numbers is independent of our ideas and so on is to say that existence claims about numbers can be justified without any appeal to our ideas. My argument, however, was directed primarily at texts written before *Basic Laws*. Is there any reason to read Frege's later claim differently? If Frege had indeed changed his mind about what it is to exist independently of our ideas, this change is not reflected in any substantive way in *Basic Laws*. The version of Begriffsschrift in *Basic Laws* is designed in such a way that the justification of existence claims for numbers can be given by logical laws alone. Basic Law V is needed, but no assumptions about our ideas or mental life are needed. Nor does Frege talk about such issues when he introduces Basic Law V or the notation for the extension of a concept. If Frege's views *had* changed, the proofs of existence claims from the laws of logic alone would not

---

passages Resnik quotes are applicable to *all* words, it is difficult to see how Resnik could use these passages in support of the attribution to Frege of this distinction between numerals and names of people. Frege does not write about ostension or baptism, and I know of no grounds on which to attribute to him a view on which some primitive notion of ostension or baptism has anything to do with his notion of Bedeutung. Dummett, who has said that some notion of ostension must underlie Frege views, also says, in *The Interpretation of Frege's Philosophy*, "This appeal to ostension was not put forward as reproducing anything in Frege, but as a possible answer to the legitimate question how the notion of the bearer of a name is itself to be explained" (159).

[26]See, for instance, Resnik, *Frege and the Philosophy of Mathematics*, 169.

suffice to show that numbers exist independently of the mind. Yet, on Frege's view, his proofs did show this.

It may seem, however, that Frege's talk of domains is more significant. Consider the following passage. "For me there is a domain of what is objective, which is distinct from that of what is actual, whereas the psychological logicians without ado take what is not actual to be subjective" (BLA 15–16). One might suspect that, if there is a separate non-physical domain of numbers, then the truths of arithmetic are truths about this special domain. For Frege to take such a position in *Basic Laws*, however, would contradict the view he took himself to be establishing in that work, the view that arithmetic is not a special science but, rather, a consequence of the most general science of all, the fundamental laws of reason alone. Even in his latest writings about the status of arithmetic, Frege's assertions that numbers are objects are accompanied by discussions of the logical structure of sentences in which number words appear. In a 1924 manuscript he talks about the oddness of attributing physical properties to numbers. He then says, "We may seek to discover something about numbers themselves from the use we make of numerals and number-words. Numerals and number-words are used, like names of objects, as proper names" (PW 265). He adds that the assertion of "Five is a prime number" looks to be an assertion that an object (five) falls under a concept (prime number), and he infers, "By a number, then, we are to understand an object that cannot be perceived by the senses" (PW 265). Thus, the significance of Frege's attributions of objecthood to numbers seems, throughout his writings, to be a strictly logical significance.

Frege's discussion of the objectivity of thoughts, however, looks different from his discussion of the objectivity of numbers. This should not be surprising, for thoughts and numbers play very different roles in Frege's picture of science. Although the laws of logic can be regarded as the laws of thought, logic cannot be viewed as the science of thought in quite the way arithmetic can be viewed as the science of numbers. Logical laws do not express facts about thoughts in the way laws of arithmetic express facts about numbers. No names for thoughts appear in the logical laws, and variable symbols are not used in place of names for thoughts but, rather, in place of sentences that express thoughts but are names for truth-values. In particular, not only does the existence of thoughts fail to be a consequence of the laws of logic, but no combination of Frege's logical symbols can be viewed as expressing the claim that thoughts exist. Thus, the existence of thoughts does not really have the status of a scientific claim. In fact, it seems that the existence of thoughts is one of the basic assumptions that Frege uses to justify the construction of his logical notation. For his explanations of what he is doing in setting up his logical notation and expressing logical laws depend on our associating a common content with sentences. Given this much it would be inappropriate simply to assume without argument that, because both thoughts and numbers are nonphysical, objective objects, all Frege's remarks about why numbers are nonphysical, objective objects can be applied to thoughts, or the corresponding remarks about thoughts can applied to numbers.

At first glance, the case for attributing a contemporary sort of platonism to Frege is much stronger if we concentrate on his comments about thoughts. In "Thoughts," Frege seems to come very close to introducing these 'abstract objects' by analogy with physical objects. He writes,

> So the result seems to be: thoughts are neither things in the external world nor ideas.
> A third realm must be recognized. Anything belonging to this realm has it in common with ideas that it cannot be perceived by the senses, but has it in common with things that it does not need an owner so as to belong to the contents of his consciousness. (CP 363)

And, in a footnote, he writes,

> A person sees a thing, has an idea, grasps or thinks a thought. When he grasps or thinks a thought he does not create it but only comes to stand in a certain relation to what already existed—a different relation from seeing or having an idea. (CP 363)

There can be no question that Frege *is* drawing an analogy here between thoughts and physical objects. But what does this show us about Frege's views about 'abstract objects'? I have already argued that, on its own, Frege's mere assertion that something is a nonphysical, nonsubjective object does not commit him to any of the contemporary epistemological or metaphysical views customarily attached to such assertions. For, in the case of numbers, these issues are inapplicable.

It is also important to note that Frege does not draw the analogy in order to provide an introduction either to his notion of thoughts or to a notion of "abstract objects". This is *not* his introduction to the notion of thought. First, this notion has been introduced previously (in "On Sense and Meaning"). Second, the analogy is not used in "Thoughts" to introduce this notion. Frege begins, in "Thoughts," by saying that a thought is something for which truth can arise (CP 353). He then proceeds, not to draw similarities between thoughts and physical objects, but to talk about the consequences of a thought being something for which truth can arise. Among these consequences is that thoughts cannot be physical objects; nor can thoughts include all the content we associate with sentences. Frege not only fails to introduce the notion of thought by analogy with physical objects, he also fails to introduce this notion by appeal to a notion of objecthood. Thus, the content of Frege's view of thoughts as objective objects is not based on an analogy with everyday medium-sized physical objects.

Furthermore, if Frege's analogy is to be taken as evidence that he embraced a contemporary version of platonism, then the obvious question to ask is whether or not he was concerned to give an explanation of how we can interact with thoughts. At first glance Frege does seem concerned with this. He says that we interact with thoughts by grasping them. And he says that when a person grasps a thought, that person comes to stand in a relation to that thought. But he also says, "To the grasping of thoughts there must then correspond a special mental capacity, the power of thinking. In thinking we do not produce thoughts, we grasp them" (CP 368). But the power of thinking cannot be understood by drawing an analogy with sense perception. More significant, whatever the relation between us and thoughts may come to, Frege has no desire to make the nature of this relation explicit. He calls the the process of grasping a thought "mysterious" and says,

> It is enough for us that we can grasp thoughts and recognize them to be true; how this takes place is a question in its own right. It is surely enough for the chemist too that he can see, smell and taste; it is not his business to investigate how these things take place. . . . So we shall not trouble ourselves with asking how we actually think or arrive at our convictions. (PW 145)

It is Frege's opinion that holding his view does not require one to answer the sort of question whose answer is considered essential to the defense of platonism by contemporary writers. What, then, is the significance of Frege's claims that thoughts are nonsubjective, nonphysical objects?

The only consequence Frege seems to want to draw from all his comments in "Thoughts" about relations between us and thoughts and a third realm is that the content, for logical or scientific purposes, of a sentence is neither a physical object (the inscription of the sentence) nor the ideas typically associated with the sentence. His aim seems to be the defense of his departing from everyday language in his logic. For, given his antipsychologism, his reformulation of sentences in his discussions of logical structure and conceptual content is reasonable only if this content of a sentence is something objective but also distinct from the characteristics of the sentence *qua* inscription.

I have argued that Frege's attempt to prove the truths of arithmetic from logical laws alone cannot be viewed as an attempt to defend a contemporary version of platonism or realism. A large part of this argument centered around the fact that Frege's Begriffsschrift has no primitive proper names. For, unless there are primitive proper names, it is a consequence of Frege's views about the admissibility of a term for purposes of science that there is no substantive way to introduce a question about whether or not proper names hook onto objects in the world. Furthermore, Frege's almost exclusive interest, throughout his career, in logic and mathematics constitutes some evidence that Frege was not concerned with the issue of whether or not proper names hook onto objects in the world. Had Frege explicitly prohibited the introduction of any primitive proper name in the construction of a systematic science, it would have been possible to argue for this directly. For it would then be clear that, for Frege, there is no substantive way to introduce a question about whether any proper names hook onto objects in the world. But no such explicit prohibition appears in Frege's writings. Nonetheless, I think it will be of interest to examine the available evidence for Frege's attitude about the permissibility of primitive proper names in empirical sciences and to consider what conclusions can be drawn.

Given Frege's lack of interest in the empirical sciences, it is not surprising that he does not explicitly address this issue. However, there are a number of reasons—other than the possibility of rendering fictional a systematic science whose primitive terms include proper names—for thinking Frege would find it inappropriate to employ primitive proper names as building blocks for a scientific field. One reason has to do with the general nature of scientific research. Although particular facts about particular objects are employed in, or justified by, scientific research, Frege seems to regard scientific fields as being marked off by the general laws of the field and to consider the purpose of introducing technical terms to be aiding in the expression of these laws. He says:

What does matter is that the word should be as appropriate a vehicle as possible for use in expressing the laws. Provided there is no loss of rigour, the more compendious the formulation of the whole system of laws is, the more felicitous is the apparatus of technical terms. (PW 5)

And:

The technical language of any science must conform to a single standard and must be judged with that standard in mind: does it enable the lawfulness of nature to be expressed as simply as possible and at the same time with perfect precision? (CP 133)

He says, about laws, "The laws of nature are general features of what happens in nature and occurrences in nature are always in accordance with them" (CP 351). And: "By its very nature, a law of nature requires universality, and reference to particular objects conflicts with this requirement" (CP 124). If the primitive technical terms of a field of science are simply the terms necessary for the expression of laws of nature, then it seems unlikely that any primitive technical term can be a proper name. This suggests that no proper names will be fundamental building blocks of any science. That Frege thinks proper names are not to figure among the primitive expressions of a science is also suggested when, in distinguishing simple from definable terms, he characterizes the defined terms as those that can be "analysed into concepts" (PW 271), for this suggests that the primitive terms that are to be used in definitions will be concept-expressions. This is also suggested by his use of the expression 'what cannot be conceptually grasped' (CP 357) for the constituents of a sentence which do not contribute to its significance for an exact science. It may also be significant that Frege's early term for the content of a sentence is 'conceptual content' and that the title of his logical notation, which is supposed to be a language adequate for the expression of what is relevant for inference and truth, is Begriffsschrift (conceptscript).

A different reason for thinking that Frege did not recognize any role for primitive proper names has to do with his rather offhand treatment of actual proper names (e.g., 'Gottlob Frege'). The expressions that Frege categorizes as proper names include, in addition to actual proper names, other object-expressions and sentences. With the exception of actual proper names, these expressions are clearly complex. If there are any primitive proper names, one would expect them to be actual proper names. But Frege says very little about actual proper names. There are only two explicit discussions of the senses of actual proper names in Frege's corpus. These appear in "On Sense and Meaning" and "Thoughts."

The discussion in "On Sense and Meaning" appears in a footnote. He says:

> In the case of an actual proper name such as 'Aristotle' opinions as to the sense may differ. It might, for instance, be taken to be the following: the pupil of Plato and teacher of Alexander the Great. Anybody who does this will attach another sense to the sentence 'Aristotle was born in Stagira' than will a man who takes as the sense of the name: the teacher of Alexander the Great who was born in Stagira. So long as the thing meant remains the same, such variations of sense may be tolerated, although they are to be avoided in the theoretical structure of a demonstrative science and ought not to occur in a perfect language. (CP 158)

It is interesting that the only modes of presentation of Aristotle which Frege considers, when he talks about the sense of 'Aristotle', are constructed from concepts that hold only of Aristotle. This is not to say that the senses of proper names are concepts. Frege is notoriously vague about what senses are, and some of his attempts at clarity seem to involve him in contradictions.[27] But the suggestion of the footnote quoted above is that the sense of a proper name is the sense of some description which picks out the object named.

Frege's other discussion of senses of actual proper names, in "Thoughts," has a similar character. He considers the possibility that two people associate different thoughts with the sentence "Dr. Gustav Lauben was wounded" because they know different things about Dr. Gustav Lauben and hence use different properties to pick him out (the doctor who is the only doctor living in a particular house; the only person who was born on 13 September 1875 in a particular town). In this case, Frege says, as far as the proper name 'Dr. Gustav Lauben' is concerned, the two people "do not speak the same language, although they do in fact refer to the same man with this name" (CP 359). Once again, the suggestion is that the sense of the proper name is the sense of a description that picks out the object named. If the sense of a proper name is the sense of a description, then the proper name can be defined, using this description.

But these actual proper names may not be the most plausible candi-

---

[27]See footnote 29 in chapter 3.

dates for primitive proper names. The most plausible candidate may seem to be a proper name introduced via use of a demonstrative pronoun, for example, 'This is Fred'. On the other hand, there is evidence that Frege thinks the use of a demonstrative pronoun can be replaced by an appropriate description. He says, in "Thoughts,"

> The case is the same with words like 'here' and 'there'. In all such cases the mere wording, as it can be preserved in writing, is not the complete expression of the thought; the knowledge of certain conditions accompanying the utterance, which are used as means of expressing the thought, is needed for us to grasp the thought correctly. Pointing the finger, hand gestures, glances may belong here too. (CP 358)

If all the relevant conditions can be spelled out, the demonstrative pronoun can be replaced by an explicit description, and the proper name introduced will not be primitive. Might Frege, nonetheless, have acknowledged the existence of cases in which it is impossible to spell out all the relevant conditions? This seems unlikely for several reasons. First, the passage quoted above immediately follows a discussion of sentences in which a time (or timelessness) is indicated by the tense of a verb along with the time of utterance. This sort of sentence constitutes the case that, Frege is saying, is the same with words like 'here' and 'there'. And it seems clear that Frege thinks that, in the case when what is left out of the everyday sentence is an explicit time-specification, the thought can be expressed by a sentence. He says, "Only a sentence with the time-specification filled out, a sentence complete in every respect, expresses a thought" (CP 370). Frege also seems to suggest that a sentence that does not express a complete thought can be modified so that it does express a complete thought when he says, "Here we must suppose that these words by themselves do not contain the thought in its entirety; that we must gather from the circumstances in which they are uttered how to supplement them so as to get a complete thought" (CP 375). Second, if some of these sentences cannot be supplemented so as to express a thought completely, then there are thoughts that cannot be completely expressed by a sentence. Given Frege's view that what is objective is what is expressible in words (FA 35), he is unlikely to acknowledge the existence of thoughts that cannot be expressed in words.

Finally, it is difficult to imagine that Frege would have accepted, as the introduction of a primitive term, the sort of event which occasionally occurs in philosophical discussions about how words come to have reference—the raising of a beer glass accompanied by the utterance of a sentence like 'This is Fred'. For the gesture and the utterance, on their own, do not pick out an object. The same sentence uttered with the same gesture might, in the context of teaching someone about table manners or the physiology of the hand, result in the assignment of a different object to the proper name 'Fred'. An elucidation of the meaning of 'Fred' requires at least some description of the context. We at least need to know the sort of object with which the speaker is concerned (e.g., medium-sized physical objects, beer glasses, gestures). But given such information, it does not seem unreasonable to suppose that on Frege's view the assignment of the intended sense to the word 'Fred' by the utterance 'This is Fred' could be—and should be—replaced not with an elucidation but with something like 'The one and only beer glass being raised now is Fred', which, presuming the concepts are meaningful and the time-specification spelled out, should be able to be taken for a definition. This requires, of course, that Frege be committed to the assumptions that the time-specification can be spelled out and that either definitions or elucidations of the concepts involved can be provided. But Frege does assume that time-specifications can be expressed and, if the concepts involved cannot be defined or elucidated, then we cannot be expected to know what 'Fred' is.

The introduction of proper names via concepts seems to be precisely what Frege requires in order to prohibit the possibility that different people associate senses with a proper name. In his discussions of the proper names 'Aristotle' and 'Gustav Lauben', Frege says that this possibility should be precluded in the language of a systematic science. In order to preclude this, these names should be defined. And it seems that the use of a demonstrative, no less than the use of 'Aristotle', requires our associating it with some concepts. This suggests that proper names introduced via the use of a demonstrative pronoun cannot be regarded as simple.

Although I have strayed rather far from Frege's texts here, my purpose is not to make any claim about Frege's views on demonstratives. Rather, my point is that there is no obvious reason to suppose that Frege thought it might be permissible to use primitive proper names in setting up systematic sciences. That Frege says so little about actual proper names, as well as his tendency, when discussing the sense of an actual proper name, to bring in descriptions of the object named, suggests that there is no way to introduce a proper name other than via concept-expressions.

Let us examine the significance of this evidence that Frege would not have permitted primitive proper names even in a language for an empirical science. First, suppose for the moment that Frege would not have permitted primitive proper names. I have argued that, with respect to numbers and mathematics, the contemporary issues raised about reference and realism cannot be raised for Frege. For Frege, no philosophical argument is required or can be given to support the claim that the truth or falsity of mathematical statements and the existence of numbers is independent from us and our minds. This claim of independence can be established only by proofs in a truly systematic science that can replace arithmetic. There is, from Frege's perspective, no general extramathematical answer to philosophical questions like "How can we have knowledge of these truths which are independent of us?" The only answer to a question about how we know that a particular statement of arithmetic is true is a proof in a systematic science. I have also argued that this characterization of arithmetic and what questions can be raised about arithmetic follow from the absence of primitive proper names in the language for arithmetic. If no logically perfect language for a systematic science ever contains primitive proper names, similar morals can be drawn for empirical science. Questions about whether some proper name for some physical object has reference amount to nothing more than an everyday scientific question that has nothing whatever to do with language.

Thus, far from introducing questions about language as a central philosophical subject matter, Frege makes the point that epistemological demands and the desire for precision should lead scientists and philosophers to replace scientific language with a logically perfect language, a language for which there are no questions to raise.

The above argument, of course, depends on an assumption for which I have no conclusive defense. This assumption is that Frege would not have permitted primitive proper names in a logically perfect language. Suppose this is in fact false. What would the consequences be for our interpretation of Frege's views? Frege never explicitly says that primitive proper names are permissible in empirical science. But this itself suggests that Frege was not primarily interested in language. Were Frege interested in a general philosophical question of how words hook onto an independent reality, the important differences between a language for mathematics and a language for an empirical science would seem to require that he at least investigate the consequences of these differences for the enterprise of making sure that all proper names in a language for an empirical science stand for objects. Yet Frege never undertook such an investigation. Furthermore, for primitive concept-expressions, Frege's descriptions of elucidation seem to say all that can be or need be said about their success in describing an independent reality. But these descriptions fall far short of establishing conditions needed for a primitive proper name to hook onto some object in an independent reality. If, on Frege's view, elucidation is the means by which a primitive proper name comes to stand for an object, then it seems to follow that the relation between a proper name in an empirical science and the object it stands for, far from being an appropriate subject for philosophical theorizing, is ultimately ineffable.

PART III

CHAPTER 6

Elucidations

THERE is a marked similarity between the conclusions I have drawn in chapter 5 about the admissibility of terms for purposes of science and the conclusions I drew earlier about Frege's view of logic. In chapter 2, I argued that Frege's view of the nature of logical laws precludes the existence of a substantive metaperspective for logic. Not only would Frege refuse to regard any metatheoretic discussion or exhibition of truth-tables as constituting a justification of a primitive logical law, he would refuse to regard any metatheoretic reasoning about primitive logical laws as expressing an objective inference. The only justification Frege would recognize for a logical law is a proof from another logical law. In chapter 5 I discussed Frege's comments about the conditions for the admissibility of a term for scientific purposes. The claim that a defined term satisfies his criteria for admissibility requires no metatheoretic argument. To demonstrate the admissibility of a defined term, it suffices to exhibit its definition. Given Frege's characterization of admissibility, anyone who understands Begriffsschrift notation will be able to see that the defined term is admissible by looking at its definition, just as she or he will be able to see that a primitive logical law is justified (and is a logical law) by looking at its Begriffsschrift expression. As for the admissibility of a primitive term, this can be shown only by using figurative expressions and guesswork to induce mutual understanding. There is no way to give a metatheoretic proof that a primitive term is admissible, for there is no way at all to guarantee that a primitive term is admissible.

There is, however, something very odd about these conclusions. In *Basic Laws* Frege does appear give arguments, in the words of everyday language rather than in symbols of Begriffsschrift, that his primitive logical laws are true, that the primitive terms of Begriffsschrift have Bedeutung, and that terms of Begriffsschrift appropriately defined from primitive terms have Bedeutung. These arguments cannot be expressed in Begriffsschrift. In fact, the very introduction of criteria for the admissibility of a term or of a logical notation seems to presuppose a metaperspective. After all, this introduction is not, and cannot be, stated in the language of Begriffsschrift. With the exception of the proofs of *Begriffsschrift* and *Basic Laws*, the views expressed in Frege's writings cannot be expressed in Begriffsschrift. The bulk of Frege's writings, then, seem to consist primarily of claims and arguments from the standpoint of some metaperspective that does not exist. So what can Frege mean to be doing in these writings?

Perhaps the most reasonable answer is that Frege's philosophical work is philosophy, not science, and of course does not constitute a contribution to a systematic science. But to stop with such an answer is to refuse to take Frege's philosophical work seriously. It is true that, for Frege, thoughts, even thoughts that are imperfect from the point of view of science, are objective. The content of thoughts, even of imperfect thoughts, can be communicated. But Frege holds that imperfect thoughts have no truth-values. In order to use a sentence to express some objective truth, the terms must be precisely defined; the thought expressed must not be one of the imperfect everyday thoughts but, rather, a thought that has a truth-value. A thought has a truth-value only if that thought can be expressed in the words of a systematic science. If what is expressed in the sentences of Frege's philosophical writings cannot be expressed in the words of a systematic science, his sentences may express something that can be communicated, but they cannot express objective truths or falsehoods.

It is a significant body of writing whose status is at stake here. There are the early sections of *Begriffsschrift*, the early sections of each volume of *Basic Laws*, and a group of sections in *Basic Laws* which alternate with sections in which proofs are given solely in Begriffsschrift symbols, all of which appear to consist of explanations or justifications of the work done in the language of Frege's systematic science. The justifications, in particular, look to be metatheoretic proofs. Then there is *Foundations* as well as most of Frege's published articles. This writing looks to consist of the explication of philosophical views, many of which have consequences for the mathematical work of *Begriffsschrift* and *Basic Laws* but which are not oriented solely toward its explanation or defense. These include, for instance, what look to be views about the expressibility of everyday language in "On Sense and Meaning." Yet my aim in this final chapter will be to argue that it follows from Frege's general epistemological views that his discursive work has the status of elucidation rather than of objective statements of facts.

Not only is the conclusion I will be drawing in this chapter controversial, it also does not appear explicitly in Frege's writings. Thus, it is important to be clear about exactly what views I am and am not attributing to Frege, the historical person. I have no qualms about claiming that Frege, the historical person, had no pretension to be offering a metaperspective for logic. I will argue that there is evidence in Frege's own writing that he did not regard the discursive sections in *Basic Laws* which appear to contain metatheoretic proofs as containing proofs at all. On the other hand, there is no direct evidence that Frege regarded the views that we regard as his fundamental philosophical views as merely elucidatory hinting at how his mathematical or logical work is to be understood. My argument that Frege's philosophical papers must, given his views, have this elucidatory status is based partly on the interpretation I have given in the previous chapters and partly on the fact that such a reading of Frege's philosophical work is needed to resolve some of the tensions in that work. In particular, I will argue that this reading not only resolves the tension that has been given the title the "concept *horse* problem" but, more significant, explains Frege's rather odd response to the concept *horse* problem.

In order to investigate the sense in which some of Frege's writings might be regarded as elucidatory, it will be important to look again at the passages in which he describes the role and the nature of elucidation. I will use these passages, as well as passages in *Basic Laws*, to support my contention that Frege did not regard the discursive sections from *Basic Laws* as presenting proofs in a metatheory but assigned them only an elucidatory role. These arguments, except where explicit disclaimers appear, are meant to concern the interpretation of the actual views in Frege's texts. The next part of the chapter is more speculative. If Frege is setting out a philosophical theory, the views expressed in some of his writings, particularly those in which he responds to the "concept *horse* problem", seem both puzzling and wrong. I will argue that these views look perfectly reasonable if Frege's central philosophical pronouncements are regarded as elucidatory. The final discussion

concerns the consequences of taking his philosophical work as elucidatory, given Frege's view of objectivity.

According to Frege, elucidation plays a role in the construction of a
systematic science. One of the requirements for setting up a systematic
science, a science whose inferences are gapless and in which it is not
possible for presuppositions to sneak in unnoticed, is the achievement
of a certain measure of precision. The statements of a systematic science
must be expressed in a logically perfect language. All terms of a logically
perfect language must either be primitive terms or be precisely defined
from primitive terms. But in every logically perfect language for a
systematic science (including the most general science of all, logic itself)
there will be primitive terms that are not and cannot be introduced by
definition. Frege says:

> Science needs technical terms that have precise and fixed meanings, and
> in order to come to an understanding about these meanings and exclude
> possible misunderstandings, we give examples illustrating their use. Of
> course in so doing we have again to use ordinary words, and these may
> display defects similar to those which the examples are intended to re-
> move. So it seems that we shall then have to do the same thing over again,
> providing new examples. Theoretically one will never really achieve one's
> goal in this way. In practice, however, we do manage to come to an
> understanding about the meaning of words. Of course we have to be able
> to count on a meeting of minds, on others guessing what we have in
> mind. But all this precedes the construction of a system and does not
> belong within a system. In constructing a system it must be assumed that
> the words have precise meaning and that we know what they are. (PW
> 207)

He gives a similar description of the presystematic role of elucidation in
science when he says,

> It is this, therefore, that serves the purpose of mutual understanding
> among investigators, as well as of the communication of the science to
> others. We may relegate it to a propaedeutic. It has no place in the system
> of science; in the latter, no conclusions are based on it. (CP 300–301)

And he goes on, in the same passage, to say that, for the success of elucidation or hinting,

> We must be able to count on a little goodwill and cooperative understand-
> ing, even guessing; for frequently we cannot do without a figurative
> mode of expression. (CP 301)

And:

> Since mutual cooperation in a science is impossible without mutual understanding of the investigators, we must have confidence that such an understanding can be reached through explication [*Erläuterung*], although theoretically the contrary is not excluded. (CP 301/KS 288)

And:

> Unlike definitions, such elucidatory propositions cannot be used in proofs because they lack the necessary precision, which is why I should like to refer them to the antechamber, as I said above. (COR 37)

Thus, although Frege places strict requirements on what can count as an admissible definition for purposes of science, the success of these definitions in eliminating any vagueness or ambiguity depends on something that lacks the precision Frege is after. Our understanding of a defined term will depend on elucidations of the content of the primitive terms that appear in the definition. But given the imprecision inherent in these elucidations, what are the grounds of his confidence in their success?

One possibility has to do with the sort of primitive terms which are admissible. Frege requires definition where definition is possible. One of his introductions to the notion of elucidation begins as follows:

> My opinion is this: We must admit logically primitive elements that are indefinable. Even here there seems to be a need to make sure that we designate the same thing by the same sign (word). Once the investigators have come to an understanding about the primitive elements and their designations, agreement about what is logically composite is easily reached by means of definition. Since definitions are not possible for primitive elements, something else must enter in. (CP 300)

Similar remarks appear in Frege's discussion of the notion of concept. He says, "What is simple cannot be decomposed and what is logically simple cannot have a proper definition"[1] (CP 182). He adds, "On the

---

[1]Actually, as G. P. Baker and P. M. S. Hacker have pointed out to me in private correspondence, this passage alone does not suffice to show that Frege thought the notion of concept was logically simple. For he goes on to say: "If something has been discovered that is simple, or at least must count as simple for the time being . . ." (CP 182–183). Baker and Hacker infer that the notion of concept is not really simple and that Frege only wanted to treat it as simple in "On Concept and Object." They claim that Frege actually defined

introduction of a name for something logically simple, a definition is not possible; there is nothing for it but to lead the reader or hearer by means of hints [*Winke*] to understand the word as is intended" (CP 183). This suggests that the introduction of a primitive term via elucidation is permissible only when, in the presystematic analysis and sharpening of the concepts of some field of science, we come upon some logically simple element. Primitive terms, then, will be terms for logical simples.

One might suspect that Frege's confidence in the success of elucidation is based either on a metaphysical view about logical simples and our access to them or on a view about how the possibility of communication is based on a shared understanding of certain logical simples. Although I think there are conclusive reasons for denying that Frege viewed external reality as composed of a set of absolute logical simples, I will not give the full argument here. For my conclusions in this chapter do not depend on refuting this view, and the inclusion of the detailed textual argument an adequate refutation requires would constitute too large a digression. I have included such an argument in Appendix B. There are indications in Frege's writings that the success of elucidation is a result of the fact that elucidation is used to coin terms for elements that underlie all communication.

One example of a Fregean elucidation is Frege's attempt to fix the meaning of the conditional-stroke (the symbol for conditional) in section 12 of *Basic Laws*. He introduces his symbol with the following remark:

'concept' both in "Function and Concept" (CP 146) and in *Basic Laws* (BLA 36). In both passages Frege says that a concept is a function whose value is always a truth-value. In neither passage, however, does Frege claim to be giving a definition. Furthermore, if this is treated as a definition, many of Frege's other writings become unintelligible. The purpose of "On Concept and Object," for instance, is to explain his notion of concept. It seems especially odd that, although "Function and Concept" was delivered before the publication of "On Concept and Object," Frege should claim, in "On Concept and Object," that the notion of concept "*must* count as simple for the time being." In particular, if his answer to Kerry can only be made out by ignoring the true definition of 'concept' and viewing the term as logically simple, then it cannot be a very satisfactory answer. Also, as I have argued in chapter 3, a proper definition must satisfy stringent criteria. Frege writes with scorn about things that look like definitions in mathematicians' writings but are not. He compares them to decorative stucco embellishments on buildings and says, "We can recognize such definitions by the fact that no use is made of them, that no proof ever draws upon them" (PW 212). The claim that concepts are functions whose values, for all objects, are truth-values is, of course, not used in any proof. In view of this and in view of the fact that he never explicitly claims to have defined 'concept', it seems only appropriate to regard the claim that concepts are functions whose values are truth-values as an elucidatory proposition rather than a definition.

In order to enable us to designate the subordination of a concept under a concept, and other important relations, I introduce the function of two arguments

$$\tau_\zeta^\xi$$

by stipulating that its value shall be the False if the True be taken as ζ-argument and any object other than the True be taken as ξ- argument, and that in all other cases the value of the function shall be the True. (BLA 51)

What, exactly is this elucidation supposed to do, and what are the grounds of Frege's confidence that it will be successful? There is an obvious and clearly inappropriate explanation of its success. Frege has given apparently mechanical criteria for determining, whenever the variables in the above expression are replaced by admissible proper names, the truth-value of the resulting expression. The mechanical nature of Frege's criteria seems sufficient to guarantee that we will be able to evaluate the truth or falsity of any sentence in which the conditional-stroke appears.

But something is missing from this account. For, on it, the content of the elucidation is presented as an arbitrary and precise stipulation. Yet the conditional-stroke is a primitive term. What is missing from the account is an account of how the purpose of elucidation—the connection of the sign with something logically simple—is affected by the passage quoted above. For Frege's point is not simply to introduce an arbitrary symbol. The conditional-stroke is supposed to be used to express something that already plays a role in presystematic thought, something his readers already understand. If he has succeeded, then his readers should recognize that his use of the precise symbol captures what is relevant to inference in the content of our imprecise everyday expressions (e.g., 'if . . . then . . .'), and his readers should be willing to use this symbol to replace the everyday expressions in scientific contexts. In the same section he goes on to show how the conditional-stroke is to be used to express the everyday use of 'and', 'neither-nor' and 'or'. The elucidation, then, does not consist of the stipulation alone but also of the connection of this stipulation with our use of everyday expressions. If this is what Frege's elucidation is supposed to achieve, why does he think it will be successful?

Frege does not say a great deal about the relation of the conditional-stroke to everyday expressions in section 12 of *Basic Laws*. Thus, it is difficult to see from an examination of this section alone why he thinks his elucidation will succeed. However, one of Frege's central tasks in promoting his logic is to communicate, via elucidation, the expressive power of his notation—that is, his understanding of the logical symbols he introduces, the primitive elements of his logic. This elucidation appears throughout his writings. Consider, for instance, Frege's discussion in *Foundations* of 'All whales are mammals'. He claims that this sentence expresses something about concepts; that what is asserted is the subordination of one concept to another. In his section 12 introduction of the conditional-stroke, Frege tells us that one of his purposes is to enable us to designate this relation of subordination in his Begriffsschrift. And in section 13 he goes on to make a similar statement about 'All square roots of 1 are fourth roots of 1'. Frege's claim that 'All whales are mammals' or 'All square roots of 1 are fourth roots of 1' is about concepts, another elucidation, can be cashed out by observing that the thought expressed can be expressed in Begriffsschrift notation by something of the form $(x)(Wx \rightarrow Mx)$.[2]

There are two obvious objections to what I have said above. First, even if Frege's introduction is supposed to connect his conditional-stroke with something logically simple, there looks to be a definition as well as an elucidation. After all, there is no imprecision in this introduction—Frege provides a mechanical criterion for determining the truth-value of every sentence in which his conditional-stroke appears. It seems that Frege is really defining a term and then arguing that this definition captures a logical notion his readers antecedently understand. Second, it is not clear how Frege's claim about 'All whales are mammals' can be regarded as an elucidation. After all, elucidations are supposed to fix the meaning of primitive terms or assign a name to some logically simple element of a science. But the claim about 'All whales are mammals' appears to do neither.

Why is Frege's introduction of the conditional-stroke not a definition? One of the hallmarks of a definition is that it must be usable in proofs. The use of a definition in a proof is either the introduction of a symbol into proof as an abbreviation of a complex expression or the removal of the symbol in favor of a complex expression that does not contain it. It is

---

[2]Of course, '$(x)(Wx \rightarrow Mx)$' probably does not really exhibit the logical form of what is expressed by 'All whales are mammals'. For unless there is some science for which the concepts of being a whale and being a mammal are to be regarded as logically simple, the legitimacy of the use of simple signs for them in a Begriffsschrift expression depends on their having been antecedently defined. If they have been defined, then the above expression does not exhibit all the logical constituents of the claim that all whales are mammals. In any case, however, the Begriffsschrift expression of the claim will have the form of a generalized conditional.

quite obvious that the introduction of the conditional-stroke cannot appear, as stated in the above quotation, in proofs stated in Begriffs-schrift notation, for Frege's description of the meaning of the conditional-stroke is stated in everyday language. In order to use this description of the conditional-stroke in the systematic science of logic, the informal terms that play a role in its logical structure (e.g., 'if', 'and') must be replaced by precise Begriffsschrift expressions. But these precise Begriffsschrift expressions will contain conditional-strokes. Thus, conditional-strokes cannot be introduced into or removed from some proof by use of a definition.

It may seem appropriate to respond that the definition is usable not in proofs of the systematic science of logic but in proofs of an extra-systematic metatheory for logic. But Frege recognizes no such use for definitions. He says, "A definition is a constituent of the system of a science" (CP 302). The role of definitions in a system is described in "Logic in Mathematics." He says that sometimes in constructing a system a combination of signs occurs over and over, and it may be of use to replace this combination with a simple sign. And he says,

> Now when a simple sign is thus introduced to replace a group of signs, such a stipulation is a definition. The simple sign thereby acquires a sense which is the same as that of the group of signs. Definitions are not absolutely essential to a system. We could make do with the original group of signs. (PW 208)

Given Frege's notion of definition, it should be clear that the introduction of the conditional-stroke is not accomplished via a definition.

What about the status of Frege's *Foundations* claim about 'All whales are mammals'? The problem with regarding this claim as elucidatory is that Frege's remarks about elucidation seem to indicate that the purpose of elucidation is to assign a sense to a primitive term.[3] If the paradigm for elucidation is the sort of introductory remark which appears in geometry texts—such claims as that a point has no extension and a line extends infinitely far in one direction—Frege's claim does not look like an elucidation. These introductory remarks contain imprecise descriptions of the sorts of things which satisfy some primitive concept, and

[3] It seems pretty clear that Frege thinks elucidations are to secure for a term both Bedeutung and sense. In the passages about elucidation he talks sometimes of Bedeutung (PW 207, CP 300), sometimes of a common understanding (CP 183, 301), and sometimes of sense (CP 301). I have chosen to talk of fixing sense here because, presuming the sense is appropriate for purposes of science, this also amounts to fixing Bedeutung.

the purpose of including such remarks in an introduction to geometry appears to be to get the reader to construct some sort of mental image of pointhood or linehood. But Frege's claim about 'All whales are mammals' does not fit this paradigm. It is not directed at such aims and does not even appear to be about the content of any primitive term. The problem with this paradigm is that much of its character comes from the synthetic a priori status of Euclidean geometry. Even Frege's uncontroversially elucidatory passages also do not appear to fit the paradigm. Consider, again, the section 12 introduction of the conditional-stroke.

It is clear from the wording of the passage in which this symbol is introduced that Frege does not mean to conjure up of mental image in his readers' minds. Nor does he mean to be giving some description of the Bedeutung of a primitive term abstracted from the role of that term in the expression of truths. Frege gives us the sense that is to be expressed by the conditional-stroke by indicating which sentences containing this term express truths and which express falsehoods. What can be said about the Bedeutung of this primitive term amounts simply to the thoughts in whose expression the primitive term occurs. Thus, the purpose of this elucidation can also be described as the communication of how this term can be used in the expression of thoughts that are true or false. Furthermore, Frege means to be communicating how to understand the thoughts in whose expression this primitive term figures by looking at how everyday sentences can be expressed in Begriffsschrift using this term. If this sort of description counts as elucidatory, then Frege's claim about 'All whales are mammals' can surely be viewed as elucidatory.

As I noted earlier, the upshot of Frege's claim is that 'All whales are mammals' is to be translated by something of the form $(x)(Wx \to Mx)$. But even if this is the sort of claim which can be elucidatory, it is still important to see *how* this claim is playing an elucidatory role for Frege. It is not difficult. In *Foundations* Frege is still trying to get his readers to understand and use his logically perfect language. The claim about the translation of 'All whales are mammals' is also introduced as part of Frege's attempt to get us to understand what he means when he says that a statement of number (*Zahlangabe*) is an assertion about a concept. To say, for instance, that the king's carriage is drawn by two horses is, according to Frege, to make an assertion about the concept of being a horse that draws the king's carriage. Or, to be more explicit, to say that the king's carriage is drawn by two horses is to say something that is roughly of the form $(\exists x)(\exists y)[x \neq y \; \& \; (z)(Hz \leftrightarrow (z = x \lor z = y))]$. This can be regarded either as an insight about numbers or as an elucidation of Frege's logical symbols. For one of Frege's points is that this symbolization makes clear our understanding of what we are and are not allowed to infer from the judgement that the King's carriage is drawn by two horses. Furthermore, Frege is convinced that this provides, not a new requirement on how his audience should hereafter understand number-words but, rather, a new articulation of something that his audience should acknowledge is actually part of their antecedent understanding of the number-words. Insofar as what is new here is the articulation rather than the content associated with the word 'two', this provides an elucidation of the simple logical notions (e.g., of quantifier and conditional) for which Frege wants to coin terms.

At this point it is possible to see why Frege thought his imprecise elucidations could be successful. For part of Frege's point is that an understanding of the logically simple elements of his logic already underlies not only our reasoning but also the success of our communication with one another. His introduction of these symbols is only an introduction of a means for articulating something his readers already understand. Because we have a common understanding of correct inference, the exhibition of how the conditional-stroke is used to express particular inferences, along with our understanding of those inferences, should be sufficient to communicate the sense of this symbol. Frege's aim is to separate out and coin terms for what is relevant to logic in the antecedently understood content of an everyday sentence. Our antecedent understanding both of the content of the everyday sentence and of correct inference should enable us to understand his use of these terms. Similarly, if we begin with any other nonsystematic science and attempt to introduce primitive terms, our antecedent knowledge of the subject matter of the science along with the use introduced for these primitive terms should be sufficient to communicate their sense.

Given the above account of Frege's notion of elucidation, let us turn to the discursive passages of *Basic Laws* with an eye toward determining whether or not they are all to be regarded as elucidatory.

Some of Frege's discursive writing in *Basic Laws* is clearly meant to be elucidatory. One of Frege's correspondents, Carl Stumpf, suggested after reading *Begriffsschrift* that Frege explain his views first in ordinary language and only then present them in his notation.[4] Frege seems to

---

[4]The editors of Frege's correspondence comment: "Frege followed Stumpf's suggestion by publishing GLA [*Foundations*] before GGA [*Basic Laws*] thus dispensing at first with his

have followed this advice in his presentation of the proofs in *Basic Laws*. Part 2 of the first volume of *Basic Laws*, "Proofs of the Basic Laws of Number", begins with an explanation of his strategy for presenting proofs. The sections of part 2 are divided into those with the heading "Analysis" (*Zerlegung*) and those with the heading "Construction" (*Aufbau*). Frege explains that all his actual proofs appear in the Construction sections. But the Construction sections are preceded by Analysis sections that are designed to make the proofs easier to understand. German words and sentences appear only in Analysis sections. In fact, except for Frege's use of his symbols rather than traditional mathematical symbols, the arguments in these sections look very much like proofs in the works of other mathematicians. They are not, however, Begriffsschrift proofs. In contrast, *not one German word appears in the Construction sections*. On Frege's view, the Construction sections alone accomplish the aim of *Basic Laws*. By preceding his actual proofs with Analysis sections in which they are explained, Frege seems to have done precisely what Stumpf suggested.

It is easy to regard Frege's Analysis sections as elucidatory because it is always quite clear what is to be elucidated and because every argument discussed informally in these sections is translated into an explicit systematic proof in a later section. But not all of the discursive arguments in *Basic Laws* are later transformed into explicit systematic proofs. In particular, Frege devotes several sections, sections 28–31 of the first volume, to an informal argument that all his names have Bedeutung.

To a philosopher who was taught in an elementary logic course to regard soundness and completeness proofs as basic theorems of logic and who also regards one of Frege's achievements to be the introduction of semantic notions, it may seem obvious that this argument is

---

conceptual notation" (COR 172). Although I do want to regard *Foundations* as playing an elucidatory role for Frege, I think the evidence of the letter from Stumpf is equivocal. Frege's Begriffsschrift is meant literally as an alternative language to that used by mathematicians. A proof expressed in Begriffsschrift will contain no words from our everyday language. But the publication of *Begriffsschrift* did not establish Frege's notation as such a language in the minds of other mathematicians. Consequently, a book written entirely in Frege's notation was likely to have a smaller audience than a book written in the everyday, semitechnical language typically used by mathematicians. Thus, Stumpf's suggestion can easily be understood as a suggestion that Frege express his views in language similar to that typically used by mathematicians. But this looks more like a description of the "Analysis" sections of *Basic Laws* than a description of *Foundations*. *Foundations* does not give versions of Frege's definitions and proofs in more familiar mathematical language. Rather, in his *Foundations*, Frege gives extensive philosophical arguments for some of the central ideas involved in his *Basic Laws* definitions and proofs.

meant as a metatheoretic proof. The argument appears to have the form of an inductive proof. Section 31 of *Basic Laws* seems to present the proof for the basis case, the proof that primitive terms denote something (*etwas bedeuten*). The proof of the inductive case on this interpretation— that if a term is defined by an admissible definition from terms that have Bedeutung, then the defined term also has Bedeutung— appears in sections 28–30. The argument in section 31 seems most like proofs in contemporary metatheory. Frege says:

> Let us apply the foregoing in order to show [*um zu zeigen*] that the proper names and names of first-level functions which we can form in this way out of our simple names introduced up to now, always have a denotation [*eine Bedeutung haben*]. By what has been said, it is necessary to this end only to demonstrate [*nachzuweisen*] of our primitive terms that they denote something [*etwas bedeuten*]. (BLA 87/GGA 48)

Thus, Frege seems to be trying to give an objective scientific argument for a metatheoretical view.

Now, on my interpretation, there are several reasons why Frege should not be able to give an objective argument that his primitive terms have Bedeutung. One reason is that it should not be possible to *prove* a term denotes something, for this is a presystematic claim and proof, or objective justification, is only possible in a systematic science or in a science that can be replaced by a systematic science. The introduction of a primitive term cannot be given in a systematic language, nor can the claim that it has Bedeutung be stated in a systematic science. To give a definition of a definable term in a systematic science is not to say that the term has Bedeutung but, rather, to introduce an abbreviation. Frege's claim that all his terms have Bedeutung is not only introduced in a presystematic context, it is intrinsically presystematic. For it can never be replaced by a systematic claim. There is no way to formulate, in a language for a systematic science, the claim that a term has Bedeutung. The argument that proper formulation of a definition ensures that the defined term has Bedeutung cannot be formulated in Frege's language, nor does he attempt to formulate it in his language.

Another reason Frege should not be able conclusively to demonstrate, even by an informal argument, that a primitive term denotes something has to do with primitiveness. Were such demonstrations possible, elucidation, since it requires guesswork and cooperative understanding, would be both unnecessary and insufficient. Finally, I have argued that Frege's view of logic precludes the existence of metalogical proofs. If this is so, the argument in section 31 should not be regarded as a part of a proof, because such a proof would be a metalogical proof.

It is by no means clear, however, that the argument in sections 28–31 creates a problem for my interpretation, for, on close inspection, it is not at all clear that Frege means to be setting out a proof. It is important to note, first of all, that the structure of Frege's argument and the words used in it do not show that this argument cannot be elucidatory. There is no real need to restrict the language in which one conveys elucidations. As long as the result is, in the end, mutual understanding between investigators, there is no reason, for instance, to ban such words as 'demonstrate' or 'show' from elucidations. In fact, an elucidation might well be more convincing if it is accompanied by informal argument. Of course, the puzzle here is not simply that words like 'demonstrate' appear in a section devoted to the meanings of primitive terms. The puzzle is, rather, that Frege appears to be giving a proof.

This situation is not as puzzling as it looks, however. For there are serious obstacles to reading sections 28–31 as the presentation of a proof by induction. If Frege's explicit Begriffsschrift proofs are to be legitimate proofs, then all names that appear in them must have Bedeutung. Thus, if it can be proved that all these names have Bedeutung, the proof is of great importance and ought to be as clearly presented as possible. But not only is the supposed "proof" that all these names have Bedeutung much more informal than anything that would be likely to count as a metatheoretic proof in an elementary logic course—Frege does not explain that he is giving a proof and what the overall structure of the proof is. He does not even include an explicit appeal to an induction principle. It would be very odd for Frege, whose aim is to increase the rigor of mathematical proof and prevent any presupposition from sneaking in unnoticed, to base the legitimacy of his entire system on such a vague proof—a proof that, indeed, is based on a principle that he does not even bother to state. He also fails to give an explicit argument for the inductive case. In section 29 he answers the question "When does a name denote something?" (BLA 84). In section 30 he introduces two ways of forming new names from old names. He says, without argument, "It follows that every name formed out of denoting names does denote something" (BLA 85). If sections 28–31 are meant to contain a metatheoretic proof that all his terms have Bedeutung, then Frege is violating fundamental methodological principles that he has been urging on mathematicians throughout his career.

The difficulties with reading sections 28–31 as presenting meta-theoretic proofs are not limited to these overall considerations. In the introduction to *Basic Laws* Frege is concerned to indicate for his readers what does and what does not count as a proof in that work. And even section 31, the section that comes closest to presenting what looks like a contemporary metatheoretic proof, does not fit the description in the introduction of what counts as a proof.[5] To see this, consider Frege's comments at the beginning of the introduction to *Basic Laws* where he discusses his proofs.

Frege says, "The proofs themselves contain no words but are carried out entirely in my symbols" (BLA 1–2). There are very few of Frege's symbols in section 31. Indeed, the "proofs" in section 31 are carried out almost entirely in words.[6] Of course, given their role in introducing Begriffsschrift, it is understandable that Frege might initially be unable to express these proofs in Begriffsschrift. However, if these are real proofs, they must be expressible in Begriffsschrift, which, after all, sets out the general laws of *all* thought. One might expect that after finishing the exposition of Begriffsschrift Frege would show that these proofs were legitimate by providing later "Construction" sections in which their Begriffsschrift proofs appear. Or, if he truly regarded them as proofs but not as expressible in Begriffsschrift, one might expect that he would mention this sort of proof in his introduction. For if these are truly proofs, then they are among the proofs on which the legitimacy of his system depends, and, if so, his standards of clarity require an explicit mention of their role. In fact, he neither attempts to express these proofs in Begriffsschrift nor indicates the existence of a special sort of proof which is not expressible in Begriffsschrift.

[5] I am indebted to Michael Resnik for suggesting that section 31 creates difficulties for my view. I have also profited from a discussion of these difficulties with Hans Sluga and Philip Hugly.

[6] There is also an interesting although, for my purposes, inconclusive footnote to section 1 of *Basic Laws*. In section 1 Frege introduces a notation for talking about a function without indicating an argument. Instead of '$\Phi(x)$', Frege writes '$\Phi(\xi)$'. He then says in a footnote: "However, nothing is here stipulated for the Begriffsschrift. Rather, the 'ξ' will not occur at all in the developments of the Begriffsschrift itself; I shall use it only in the exposition of it, and in elucidations" (BLA 34). Frege uses 'ξ' in section 31. This is, of course, not conclusive in any way. For the title of part 1 of *Basic Laws* (which includes section 31) is "Exposition of Begriffsschrift." And there is no question that section 31, unlike sections 50–52 of part 1, contains no Begriffsschrift proofs. Frege's mention here of the exposition of Begriffsschrift and elucidations, however, is significant. For the exposition of his notation is mostly presystematic; its purpose is to enable us to understand and make substantive use of Begriffsschrift. Thus, the exposition itself seems to have a primarily elucidatory role.

This might be viewed as an oversight on Frege's part, but such an oversight seems unlikely. For Frege indicates that he is aware of the suspect nature of these "proofs" in the introduction to *Basic Laws*. He specifically discusses the status of sections 28–31 in the introduction. The relevant passage begins with the remark "Some matters had to be taken up in order to be able to meet all objections, but are nevertheless inessential to an understanding of the propositions of Begriffsschrift" (BLA 9). He follows this remark by suggesting that certain sections be skipped on a first reading of *Basic Laws*. Among these are sections 26 and 28–31. Of course the status of these sections is still not entirely clear, but it does seem that Frege does not regard them as parts of his actual proof (a characteristic sign of an elucidation). Frege goes on to say that, after reading the expositions of his logical system the first time through, one might go back and read the entire *Basic Laws*, and he says that this should be done "Keeping in mind that the stipulations that are not made use of later and hence seem superfluous serve to carry out the basic principle that every correctly formed name is to denote some-thing, a principle that is essential for full rigor" (BLA 9). If these are proofs, they are proofs whose conclusions are never used in later *Basic Laws* proofs.

Let us compare the role assigned to these stipulations with a passage from *Foundations* in which Frege discusses elucidations.

> When the author feels himself obliged to give a definition, yet cannot, then he tends to give at least a description of the way in which we arrive at the object or concept concerned. *These cases can easily be recognized by the fact that such explanations are never referred to again in the course or the subsequent exposition.* (FA viii, my emphasis)

Of course, the discussion in section 31 is not a psychological descrip-tion. But the psychological descriptions Frege talks about here are ex-amples of elucidations. And it is because they are elucidations rather than actual premises of proofs that they are never referred to again. This is also supported by the following passage from "Logic in Mathemat-ics":

> When we look around us at the writings of mathematicians, we come across many things which look like definitions, and are even called such, without really being definitions. . . . We can recognize such definitions by the fact that no use is made of them, that no proof ever draws upon them. (PW 212)

An explanation that is introduced but never used again can, at most, serve the purpose of eluciation. This describes the explanations in section 31 perfectly.

Let us return now to the passage from the introduction to *Basic Laws*. After claiming that sections 28–31 are inessential to an understanding of the sentences of Begriffsschrift, Frege goes on to explain the purpose of these sections.

> In this way, I believe, the suspicion that may at first be aroused by my innovations will gradually be dispelled. The reader will recognize my basic principles at no point lead to consequences that he is not himself forced to acknowledge as correct. (BLA 9)

Proofs do not dispel suspicion—proofs establish truths. If Frege can prove that his basic principles never lead to incorrect consequences, he should do so. But Frege cannot do this. In fact, were such proofs necessary, it would suggest that something is seriously wrong with Frege's basic laws. For a basic logical law, on Frege's view, must be immediately evident from the sense of its expression (CP 405). Hence, there is a role for just such an elucidation as section 31 seems to provide in getting his readers to accept his basic principles.

I have argued that, in an important sense, most of the discursive writing that seems as if it might express informal proof in *Basic Laws* should be regarded as elucidatory. In particular, I have argued that the existence of apparently metatheoretic proofs in *Basic Laws* does not provide evidence that Frege recognized the existence of a substantive scientific metaperspective for logic. In these arguments, as I indicated earlier, I do not mean to be going beyond the views of Frege, the historical person. Given Frege's standards of rigor and clarity, it seems evident that, had he regarded any of his writings as setting out objective metatheoretic proofs, he would have been explicit about the distinction between systematic and metatheoretic proof and the grounds for accepting metatheoretic proofs as establishing truths.

Of course, if any of Frege's writing is to be regarded as elucidatory, these sections of his mathematical writings would seem to be the most obvious candidates. But I suggested earlier that virtually all Frege's philosophical work should be regarded as elucidatory, and the discussions of the discursive sections of *Basic Laws* does not provide evidence for this. How can it be reasonable to take Frege's more philosophical writings to be playing a merely elucidatory role for his mathematics rather than presenting philosophical theories?

There are, first of all, reasons for thinking that, given his view of the reception of *Begriffsschrift*, Frege might have thought there were practical reasons for giving more elucidation for his mathematical work than is customary. While many of the reviews of this work were positive, virtually none of the reviewers understood the difference between Frege's logical notation and Boole's.[7] The mathematical details of *Begriffsschrift* were apparently insufficient to communicate this difference to his audience. Frege responded by writing several discursive papers in which he discussed the aim and justification of Begriffsschrift. Furthermore, as Bynum argues in "On the Life and Work of Gottlob Frege,"[8] there is some reason to believe that Frege had an easier time publishing more discursive works. Two of Frege's papers on the difference between his and Boole's notations—papers that included more of his symbols—were rejected for publication. One might suspect that more of Frege's subsequent discursive work was also molded by the reception of *Begriffsschrift*.

There is also evidence in *Basic Laws* that *Foundations*, as well as several of Frege's more famous papers that appeared after the publication of *Foundations* and before the publication of the first volume of *Basic Laws*, is meant to be doing some elucidatory work for Frege. In the early sections of *Basic Laws*, when Frege is explaining the rules for expression in his Begriffsschrift notation and the meaning of its primitive symbols, he frequently footnotes arguments or explanations from *Foundations* (BLA 29, 31) as well as from "On Sense and Meaning" (BLA 35, 38), "On Concept and Object" (BLA 31, 33, 37), and "Function and Concept" (BLA 33), the three later papers in which he introduces some of the changes he made in his logic after *Foundations* but before *Basic Laws*. Also, Frege begins part 3 of *Basic Laws*, which is meant to introduce his construction of the real numbers, with an extensive discussion of the nature of definition and the views of others. In spite of the fact that the other views he discusses are almost exclusively the views of mathematicians rather than philosophers, his discussions in part 3—especially the discussion of formalism—look as philosophical as those in *Foundations*.

In fact, since almost all Frege's writings concern mathematics or logic, it is not difficult to assign an elucidatory role to his discursive writings.

---

[7]Translations of several reviews of *Begriffsschrift* appear in *Gottlob Frege: Conceptual Notation and Related Articles*, translated and edited by Terrell Ward Bynum. Although Venn's and Schröder's reviews are very negative, most of the other reviews are positive. However, none of these reviewers seems to understand the quantifier notation.

[8]See CN 21–22. Bynum discusses the reception of *Begriffsschrift* at some length in CN 15–24.

Consider, for instance, some of the more significant discursive writing so far unmentioned, the two papers titled "On the Foundations of Geometry" and the three-part work titled "Logical Investigations" which consists of the papers "Thoughts," "Negation," and "Compound Thoughts." It is also possible to give an account of the sense in which these papers can be regarded as elucidatory. The papers on geometry are meant as an attack on Hilbert's mathematical work on the foundations of geometry. On the other hand, the papers of "Logical Investigations" look more like straightforward philosophy papers. They are not explicitly tied to a mathematical project. In particular, they are not meant as a contribution to the project that began with *Begriffsschrift* and was to have been completed with *Basic Laws*. For Frege had, by this time, given up his attempt to prove the truths of arithmetic from logical laws. But some of Frege's unpublished writings suggest that there is an elucidatory project for which he would have wanted to write something like "Logical Investigations." After Frege had given up on his original project of trying to prove the truths of arithmetic from logical laws, he seems to have decided that his main contribution was his logical notation. His unpublished writings contain a great number of attempts at introductions to logic texts. Many of the claims that appear in these unpublished introductions also appear in the three papers that make up "Logical Investigations." Thus, these papers can be viewed as a defense of Frege's basic logical notions, as one more contribution to his battle to establish the superiority of his logic and his notation over those of others.

But if it is *possible* to find an elucidatory purpose, in the manner sketched above, for each of Frege's apparently philosophical works, it is also not entirely clear why this is *desirable*. For the extent and depth of the philosophical discussion these works contain far exceeds that of most works of mathematics or mathematical texts. These papers can rightly be viewed as contributions—major contributions—to philosophy. In suggesting that these writings should be viewed as elucidatory, I do not mean to be suggesting otherwise. It seems clear that Frege did mean his work as a contribution both to philosophy and to mathematics and that he thought, in particular, that mathematicians ought to pay more serious attention to philosophy. I mean to be suggesting that on Frege's view there are important differences between the enterprise of philosophy and the enterprise of mathematics or science—that the purpose of philosophical work, unlike that of scientific work, is not objective theorizing or the establishment of truths.

I will end this chapter with a discussion of the significance of Frege's having tied his notion of objectivity to his standards of precision and rigor of expression. Before I turn to this discussion, however, I want to give one more argument for thinking his philosophical writing must have the status of elucidation. There is one paper that looks far too philosophical to be of interest to most mathematicians and yet whose views, on close inspection, make sense only if we view Frege's aim as elucidatory. This paper is "On Concept and Object," his response to Benno Kerry's objections to the Fregean notion of concepthood. I will argue that, if we view the purpose of this paper as elucidatory, we can explain why, on Frege's view, Kerry's objection really was insignificant and appropriately dismissible as an "awkwardness of language."

Between 1885 and 1891, Benno Kerry wrote a series of articles entitled "Über Anschauung und ihre psychische Verarbeitung" which included a now-famous objection to Frege's use of 'concept'. This objection has come to be known as the 'concept *horse* problem'. And it is commonly believed that Frege's reply, in "On Concept and Object," misses the real import of Kerry's comments.[9] In fact, although Frege is widely regarded as an exceptionally clear writer, the issue of how to interpret Frege's reply has caused a great deal of confusion. I believe that Frege's reply is not mysterious and that it is both appropriate and adequate as a response to Kerry. The widespread confusion, I believe, is a consequence of an insufficient appreciation of the importance of elucidation in Frege's writing.

The explicit objection to which Frege responds amounts to a counterexample to Frege's claim that concepts cannot be objects and objects cannot be concepts. Kerry suggests we consider the statement 'The concept *horse* is a concept easily attained'. This seems to be a true statement. In it something—the concept *horse*—is said to fall under a concept, that of being a concept easily attained. What falls under a

[9]Many of the papers in E. D. Klemke, ed., *Essays on Frege*, are devoted primarily or in large part to discussions of versions of the 'concept *horse* problem'. Some more recent discussions appear in Cora Diamond, "Throwing Away the Ladder"; V. H. Dudman, "The Concept Horse"; Peter Geach, "Saying and Showing in Frege and Wittgenstein"; Philip Hugly, "The Ineffability in Frege's Logic"; Terence Parsons, "Why Frege Should Not Have Said 'The Concept *horse* is not a Concept.'" It has been noted by several of these writers that, if the notion of concept is not part of a metaphysical theory but is used only in order to aid in the explication of Begriffsschrift notation, then there is no problem. Diamond and Geach, in their articles cited above, as well as Jean van Heijenoort in "Logic as Calculus and Logic as Language," suggest either that this actually was or may have been Frege's view. The more prevalent attitude, however, is expressed by Dudman, who says that it would be "absurd" to maintain that this is how "Frege intended his discourse about concepts to be interpreted" (74).

concept must, on Frege's understanding of 'object', be an object.[10] Hence the concept *horse* must be an object. But it is also, assuming the statement in question is true, a concept easily attained. Hence the concept *horse* is also a concept.

Frege's response is to say that his use of the term 'concept' has been misunderstood. Frege contends that, as he understands 'concept', the concept *horse* is not a concept. While it may seem that the concept *horse* is a concept easily attained, this is only a symptom of the difference between our everyday use of 'concept' and his own use of this term. How could it be that, for Frege, the concept *horse* is not a concept?

Given Frege's tendency to suggest, when there is confusion over the meaning of a term, that the term should be defined, one might expect that Frege's strategy, in answering Kerry, would be to define 'concept'. In this case, however, Frege not only declines to define 'concept' but responds, "What is simple cannot be decomposed and what is logically simple cannot have a proper definition" (CP 182). It seems reasonable to infer that Frege means to take 'concept' as a primitive term of some science; if so, the science must be logic. For he also says that the notion of concept belongs to logic (CP 135) and, in the introduction to *Basic Laws*, "Concept and relation are the foundation-stones upon which I erect my structure" (BLA 32). Furthermore, part of his evidence that truths of arithmetic are truths of logic is that statements of number are statements about concepts.[11] However, 'concept' is not a primitive term of Frege's notation. The term 'concept' appears in his presystematic discussions of how Begriffsschrift notation can be used to represent the logical regimentation of thoughts. Unlike the logical notion of negation, for which Begriffsschrift has a primitive symbol, the negation-stroke, the notion of concept is not represented by any primitive Begriffsschrift symbol. Given the role Frege assigns to primitive terms, it is not clear that there is a respect in which 'concept' can be regarded as a primitive term.

[10]This may look to be a circular explanation of the notion of object. However, I do not mean to be giving an explanation of Fregean objecthood here but only to be indicating briefly the structure of Kerry's objection as Frege understands it. Later in this chapter I will discuss Frege's notions of objecthood and concepthood at greater length.

[11]A statement of number (*Zahlangabe*) is a statement in which there is said to be a certain number of a certain sort of thing (e.g., Frege had two horses). A central insight of Frege's *Foundations* is that a statement of number contains an assertion about a concept (see section 46). 'Frege had two horses', for instance, should be understood as an assertion about the concept of being a horse of Frege's. This is simply to say that its logical regimentation would be something of the form: $(\exists x)(\exists y)[x \neq y \ \& \ (z)(Hz \leftrightarrow (z = x \lor z = y))]$.

Instead of looking at Frege's use of 'concept' and 'relation' in his presystematic discussions, I will look at his use of 'function'. My reason is that, in spite of his claim that the foundation stones of his structure are concept and relation, Frege does not begin either of his introductions of his Begriffsschrift with a discussion of concept and relation. Instead, he begins with the notion of function. In *Begriffsschrift* he also precedes the discussion of the notion of function with an introduction of the conditional-stroke, the negation-stroke, and the identity sign. Functions are brought in in *Begriffsschrift* only in order to explain generality and the quantifier notation. His later division of possible contents of judgements into thought and truth-value, a consequence of the *Sinn/Bedeutung* distinction, allows him to construe all primitive terms of Begriffsschrift as function signs. Thus, in *Basic Laws* the introduction of the notion of function precedes the introduction of any actual term of Begriffsschrift. The introductions of the notion of function in both works revolve around discussions of the significance of the part of an expression which is left over when we regard one particular part of the expression as replaceable by something else.

Frege begins *Basic Laws* with a section titled "The function is unsaturated," a section containing no Begriffsschrift symbols at all. He explains that functions are not expressions, that they require completion by an argument that is saturated, and that, when a function is completed by an argument, the result is the value of the function for the argument (BLA 34). Negation, for instance, is a function whose value is the False when the True (the object for which every true assertoric sentence stands) is its argument and whose value, for any other argument, is the False. The introduction of the notion of concept, a function whose value is always a truth-value, also precedes the introduction of Begriffsschrift symbols in *Basic Laws*.

Both introductions of function and concept clearly require a sense of what sort of expression is replaceable or can be regarded as occupying an argument place.[12] In effect they are not introductions of function and

---

[12]In *Frege: Philosophy of Language*, Michael Dummett argues that Frege's notion of object is that of whatever a proper name (or singular term) stands for. Dummett also says that categories of expressions other than that of proper names "are defined inductively, starting with 'proper names' as a basis" (FPL 54), and that Frege "was content to allow the whole distinction between proper names and expressions of other kinds to depend upon intuitive recognition, guided only by the most rough and ready of tests" (FPL 54). Dummett concludes: "Such an attitude is not acceptable." It should be evident from my discussions of objecthood and concepthood and of definition that I disagree and that I

concept so much as illustrations of how what is expressed by a string of signs can be divided into constituents that have logical significance. Frege has argued that the traditional regimentation of sentences into subject and predicate does not reflect a logical regimentation. In his logical regimentation the elements are concepts and objects. The re-

---

think Frege would disagree both with Dummett's characterization of Frege's explanation of objecthood and with Dummett's conclusion.

Dummett's claim looks plausible for two reasons. First, Frege concentrates on descriptions of language when he characterizes functionhood or concepthood. Second, Frege appears to presuppose, in most of these characterizations, either an understanding of objecthood or a recognition of which terms are singular terms (or proper names). In both *Begriffsschrift* and *Basic Laws* Frege introduces the notion of function by talk of regarding certain parts of expressions as replaceable. The parts in question, of course, must be singular terms. In "On Concept and Object" Frege introduces concepthood by contrasting grammatical predicates with singular terms. Thus, it may seem that an intuitive understanding of grammar and, in particular, what counts as a singular term is the basis for Frege's distinctions.

But the understanding that Frege implicitly relies on is of logical, not grammatical, structure. In particular, given that one of the motivations of Frege's construction of Begriffsschrift is to avoid the logical imperfections of everyday language, the existence of a grammatical distinction should not, on its own, be sufficient to justify the claim that there is a logical distinction, as Frege himself says (see, e.g., PW 143). Furthermore, even in "On Concept and Object" Frege does not invariably follow grammatical form. For although 'The horse is a four-legged animal' appears to have the grammatical form of a simple predication, Frege is not loath to say that it should not be understood as a simple predication (CP 185). Also, he does not always use grammar in his characterizations. Another expression Frege uses for the predicative nature of concepts is 'unsaturated' (see, e.g., CP 187). He says elsewhere: "It is really in the realm of sense that unsaturatedness is found and it is transferred from there to the symbol" (CP 393).

I think that all of this indicates that what we need to understand, in order to see what his notion of concepthood is, is not its grammatical form at all, but rather the thoughts expressed by our sentences. Frege's view presupposes that we share a common store of thoughts and that thoughts are logically articulated. An understanding of a thought includes a grasp of correct inferences in which it can figure. But we cannot talk about particular thoughts and their constituents without using sentences that express these thoughts. Frege says: "that a thought of which we are conscious is connected in our mind with some sentence or other is for us men necessary" (PW 269). Thus, his examples are given by examining sentences and their parts.

Frege's purpose in talking about linguistic categories is, as Dummett says, to talk about logical categories. But our recognition of these categories is not a prerequisite for our recognition of the logical categories. Rather, Frege's view is that, given that we have a common science, that we do communicate and we do agree on what counts as correct inference, we implicitly recognize the logical categories. Our antecedent knowledge of what is expressed by certain sentences is used to guide us to an understanding both of the logical categories into which Frege regiments thoughts and of the linguistic categories that mirror them.

In fact, we recognize singular expressions in sentences by using our understanding of the thoughts expressed by these sentences. Frege says: "If, then, we look upon thoughts placement of subject/predicate regimentation with concept/object regimentation makes it possible for Frege to formulate a logical notation that has important advantages over Boolean or Aristotelian logical notation. Frege's notions of function and concept are also illustrated by the use of function signs and object signs in the Begriffsschrift expression of everyday claims and inferences of arithmetic. Ultimately, his readers' understanding of concept/object regimentation must come from their understanding of correct inference and how Frege's regimentation marks off constituents that play logical roles in inferences.

Although there are no symbols for the terms 'function', 'concept', and 'object' in Frege's Begriffsschrift, his introduction of these terms should not be regarded as playing the role of a stucco embellishment on a building. For an understanding of his use of these terms is required if one is to use his notation. The sorts of symbols which can be function-signs or concept-signs differ from the sorts of symbols which can be object-signs (or, proper names). Frege begins the second chapter of *Begriffsschrift* by saying

> We have already introduced a number of fundamental principles of thought in the first chapter in order to transform them into rules for the use of our signs. These rules and the laws whose transforms they are cannot be expressed in the ideography because they form its basis. (BEG 28)

---

as composed of simple parts, and take these, in turn, to correspond to the simple parts of sentences, we can understand how a few parts of sentences can go to make up a great multitude of sentences, to which, in turn, there correspond a great multitude of thoughts" (CP 390). Instead of interpreting Frege's claims about sentences and their constituents, in "On Concept and Object," as claims about language, we can interpret them as claims about thoughts and their constituents. When Frege characterizes 'The morning star is a planet' as expressing a simple predication, this can be viewed as a claim about one way of analyzing the thought expressed by this sentence. We know that this thought can be analyzed into constituents that are expressed by 'The morning star' and 'is a planet' not because of the grammatical form of the sentence but, rather, because of the legitimate inferences it can be used to express. From what is expressed by this sentence we can infer that there are planets. On the other hand, from what is expressed by 'The unicorn is a one-horned horse', which looks grammatically like a simple predication, we cannot infer that there are one-horned horses. The thought expressed by this sentence cannot be analyzed as a simple predication.

Finally, Frege's point in getting his readers to recognize the logical categories is that this will enable them both to understand his Begriffsschrift and to see its superiority for representing inferences and logical content over other logical notations. The logical categories cannot be defined and there is no reason to define the linguistic categories that mirror them. Thus, in particular, Frege is not giving an inductive definition of expressions other than proper names. Tyler Burge also argues against Dummett's objection in "Frege on Truth."

The introduction of the notion of function or concept is necessary for explaining how Frege's notation is to be used in the expression of thoughts.

Of course, it cannot suffice, as a response to Kerry, for Frege to claim that he uses 'concept' as a technical term for explicating his logical system and that Kerry has not grasped the meaning of this technical term. For if Frege's use of this term is foundational for logic, the success of that logic as a scientific tool requires that the meaning of 'concept' not be private to Frege. Frege's elucidations must actually succeed. And, for this success, it must be possible for Kerry and others to understand Frege's use of 'concept'. This must include an understanding of what Frege means when he says that the concept *horse* is not a concept.

In addition to the explanations of the notions of concept and object he offers in his mathematical works, Frege makes another attempt to achieve this mutual understanding in "On Concept and Object." He begins by saying "A concept (as I understand the word) is predicative" (CP 183). He adds, in a footnote, "It is, in fact, the meaning of a grammatical predicate." Thus, in a sense, it should be easy to see that Kerry is wrong about the concept *horse*. For, were the concept *horse* a concept, the meaning (Bedeutung) of 'the concept *horse*' would be the meaning (Bedeutung) of some grammatical predicate. What grammatical predicate would this be? The obvious first choice is the predicate 'is a horse'. But if 'is a horse' and 'the concept *horse*' have the same Bedeutung, it must be possible, whenever one of these expressions appears in a sentence, to substitute the other for it without altering the truth-value of the sentence. However, substituting one of these expressions for the other results not in a different sentences with the same truth-value, but in nonsense. The sentence

Gottlob Frege is a horse

is a perfectly legitimate, although false, sentence. But

Gottlob Frege the concept *horse*

is not a sentence at all—it is simply a string of words. Kerry could not retreat by saying that 'the concept *horse*' has the same Bedeutung as 'a horse', for the result of replacing 'a horse' in 'Gottlob Frege is a horse' with 'the concept *horse*' is

Gottlob Frege is the concept *horse*.

It is clear that if Gottlob Frege is a horse, then he is not the concept *horse*. This substitution also fails to preserve truth-values. The Bedeutung of a grammatical predicate cannot also be the Bedeutung of an object-name (proper name) and, consequently, object-names cannot name concepts.

This argument may seem too quick. Perhaps it only shows that Frege is wrong about the intersubstitutivity of terms with the same Bedeutung. But a closer examination reveals insuperable obstacles to making a proper name stand for a concept. To see this, it is important to remember that, on Frege's view, there are only two available strategies for establishing that some object-name (e.g., 'the concept *horse*') stands for a concept. Either the content of the expression in question must be obtained by splitting up a proposition (see, e.g., PW 16–17) or it must be obtained by definition from primitive terms. It should be obvious that no proper name can be shown to stand for a concept by splitting up a proposition in which the proper name appears. Proper names, by virtue of being proper names, do not play the grammatical or logical role of concept-expressions. However, it may at first not be at all obvious that we cannot make 'the concept *horse*' stand for a concept by definition. Definitions, after all, are arbitrary stipulations. Thus, it ought to be possible to define 'the concept *horse*' as anything we like—even a concept.

But this strategy will not work either. Suppose, for instance, we tried to define 'the concept *horse*' by stipulating that 'the concept *horse*' will hereafter stand for the Bedeutung of 'is a horse'. Does this definition succeed in making 'the concept *horse*' stand for a concept? It succeeds only if the expression

the Bedeutung of 'is a horse'

stands for a concept. But 'the Bedeutung of "is a horse"' is—as Frege says about 'the Bedeutung of "is a square number"' (COR 142)—a proper name. The burden of this argument was to establish that it is possible for a proper name to refer to a concept. In fact, whatever is substituted for the blank space in

'The concept *horse*' stands for _____

must be a proper name. Thus, we cannot show that a proper name can stand for a concept by defining it unless we can start out with a proper name that stands for a concept.

This should not be surprising. For the difference in the logical roles of objects and concepts is mirrored in differences in the employment of proper names and concept-expressions. If a proper name, or object-expression, is deleted from a sentence that expresses a thought, what results from filling the empty space with a concept-expression not only will fail to express a thought but will not even be a proper sentence. Similarly, the difference in logical roles is mirrored by a difference in definitions of object- and concept-expressions. It is a consequence of the logical structure of definitions of object-names that it is impossible to give a definition that makes an object-expression stand for a concept. For the definition of an object-expression will be an identity, but there can be no criteria of identity for concepts. Frege writes, in his unpublished comments on Sense and Meaning, "The relation of equality, by which I understand complete coincidence, identity, can only be thought of as holding for objects" (PW 120). After all, what can be said of objects cannot be said of concepts. Yet one might suspect that there is, nonetheless, a way of talking about when concepts are or are not identical. Frege goes on to say that there is an analogous second-level relation between concepts. Two pages later, he adds:

> Now we have seen that the relation of equality between objects cannot be conceived as holding between concepts too, but that there is a corresponding relation for concepts. It follows that the word 'the same' that is used to designate the former relation between objects cannot properly be used to designate the latter relation as well. If we try to use it to do this, the only recourse we really have is to say 'the concept Φ is the same as the concept X' and in saying this we have of course named a relation between objects, where what is intended is a relation between concepts. (PW 121–122)

He repeats these views in 1906 (PW 182). It is important to be clear about the significance of these remarks. We cannot say that this corresponding second-level relation between concepts *is* the identity relation for concepts. There is no relation of identity of concepts. And since there is no relation of identity of concepts, the corresponding second-level relation cannot give us a criterion for identity of concepts. We simply cannot say that concepts are the same or different. There is only one thing of this nature that can be said of concepts. It is that their extensions are the same, or that $(x)(Fx \leftrightarrow Gx)$. Thus, on Frege's view, it is not that there are two sorts of claims—that F and G are the same and that $(x)(Fx \leftrightarrow Gx)$—which mean the same thing. Rather, there is no substantive claim that can be made about identity of concepts. On Frege's view, $(x)(Fx \leftrightarrow Gx)$ is *all that can be said*.[13]

So Frege has an answer to Kerry. A concept can be named only by a grammatical predicate, and there is no grammatical predicate that names what is named by the expression 'the concept *horse*'. As Frege says some years later, in a letter to Russell, "If we want to express ourselves precisely, our only option is to talk about words or signs" (COR 141). Kerry cannot claim to be unable to distinguish grammatical predicates from other linguistic expressions, for to do so would be to claim to be unable to use language. Kerry also cannot claim that Frege has inaccurately explained the meaning of 'concept', for Frege does not take himself to be explicating any ordinary word but to be coining a technical term. But there is still something odd about Frege's answer.

The above discussion makes use of our ability to ask what looks to be a perfectly legitimate sort of question: that is, whether something, say *A*, is a concept. And if there is any content to the question of whether *A* is a concept (rather than if '*A*' is a grammatical predicate), then '*A* is a concept' and '*A* is not a concept' should express something informative. But if we concatenate two grammatical predicates, the result is not a sentence (e.g., 'is a horse is a concept'). Since only grammatical predicates can name concepts, it follows that we cannot use the word 'concept' in order to say, of a concept, that it is a concept. We can, of course, say '*A* horse is a concept'. But this is of no help in solving the problem. For the content of this claim is not that the expression '*A* horse' stands for a concept. It is, rather that (x)(if *x* is a horse, then *x* is a concept). Whenever concatenating some expression with 'is a concept' results in the predication of concepthood, that expression must be an object-name.

Frege writes, in 1906, "The word 'concept' itself is, taken strictly, already defective, since the phrase 'is a concept' requires a proper name as grammatical subject and so, strictly speaking, it requires something contradictory" (PW 177–178). He also notes (see PW 239, COR 141) that the term 'function' is as defective as the term 'concept'. In fact, as he

---

[13]Note that, although there are no identity criteria for concepts, it does not follow that concept-expressions cannot be defined. A definition of some concept-expression, say *C*, is appropriate only if *C* has no meaning. In order to define *C* one simply gives a concept-expression that determinately holds or does not hold of every object. Such a definition might have, roughly, the following form: *x* is a *C* $\leftrightarrow$ *x* is a *D* (where *D* is a concept-expression that is appropriately constructed from primitive terms). However, it does not follow, from this, that *C* and *D* *are the same concept*.

indicates in his correspondence with Russell (COR 136), it follows from his elucidation that whenever the blank spaces in the expressions '_____ is an object' and '_____ is not a concept' are properly filled in, they express truths. It follows that 'Everything is an object' and 'Nothing is a concept' express truths. This is, in part, what Frege is talking about when he says, in "On Concept and Object" (CP 193), "My expressions, taken literally, sometimes miss my thought." But the problem is not, as this quotation might suggest, that when Frege uses these expressions he sometimes ends up not quite saying what he means. The problem is that the defects of the words 'function' and 'concept' seem to preclude making sense not only of the above discussion about the concept *horse* but of much of what Frege actually says when he uses the words 'concept' and 'function'. For instance, Frege needs to draw a distinction between concepts and objects. There is no obvious problem with Frege's saying that no concept is an object. It seems true. However, since it must also be true that there are no concepts, it also follows that all concepts *are* objects. Similarly, all concepts are horses, all concepts are cats, all concepts are unicorns. But if all concepts are horses, what are we to make of one of Frege's central claims in *Foundations*—that assertions of number are assertions about concepts?

It may seem that the appropriate response to the above observations is suggested by Frege's remark that "the words 'function' and 'concept' should properly speaking be rejected. Logically, they should be names of second-level functions" (COR 141). One might be tempted to try to remove all Frege's paradoxical remarks by replacing attributions of concepthood with second-level assertions. We can replace assertions that employ the expression 'is a concept' with assertions employing a second-level concept-expression that yields a sentence only when a concept-expression is added and that yields a true sentence whenever a concept-expression is added. An example, in everyday language, is Dummett's expression "is something which everything either is or is not" (FPL 216). Dummett goes on to suggest that the problematic sentence 'The concept *horse* is a concept' can be replaced by 'A horse is something which everything either is or is not'.

To view this as a solution to Frege's problem is to view the apparently paradoxical nature of Frege's remarks as the problem. But this really is not the problem. To see this, consider the purposes for which Frege uses the term 'concept'. This term is used for the presystematic discussions designed to explain the notion that underlies the different uses of different sorts of symbols in the Begriffsschrift expression of thoughts.

To eliminate the expression 'concept' in Frege's explanation of Begriffsschrift symbolization in favor of 'something which everything either is or is not' is not simply to make the explanation more cumbersome but to presuppose what is to be explained.

The English words involved in Dummett's reformulation do not suffice to make the distinction Frege wants to make. After all, on Frege's view the number one is an object. But there is nothing wrong with the sentence 'The number one is something which everything either is or is not'. The difference between this sentence and 'A horse is something which everything either is or is not' is that 'is' serves as an indication of predication in the latter, while in the former it stands for identity. If Dummett's second-level expression is to be used to explain concepthood, then it must be distinguished from the apparently identical first-level expression, but this distinction can be made only by use of the notion of predicativity, in other words, concepthood. The difference cannot be explained to someone who does not understand the notion of predicativity Frege is trying to explain. The only unambiguous option is to use logical notation, for example, $(x)(\ldots x \ldots \lor \sim \ldots x \ldots)$. But this is useless, given that the point of Frege's discussion of concepts is to introduce logical notation.

In effect, the only way to eliminate Frege's apparently paradoxical remarks is to resort to use of Begriffsschrift regimentation, but this cannot be done unless we already understand Begriffschrift regimentation. And the purpose of Frege's apparently paradoxical remarks is to get his readers to understand this regimentation. It seems, then, that for Frege's purposes the apparently paradoxical remarks cannot be eliminated. But although there seems to be a problem here, it is not obvious that the apparently paradoxical nature of Frege's claims about concepts is the problem. To see that this does not constitute the problem it will help to examine the sense in which his apparently nonparadoxical uses of the word 'object' also seem to miss his thought. The problem with the word 'object' is that, given Frege's notion of objecthood and concepthood, the assertion that something is an object seems empty. If everything is an object, what are we to make of the content of Frege's apparently substantive claim that numbers are objects? Frege's real problem is that it seems that the distinction between concept and object is a fundamental logical distinction, yet he can give no content-full explanation of concepthood and objecthood. That is, it seems that what Frege wants to say is unsayable. If Frege's words miss his thought and, in fact (as it seems), *must* miss his thought, what grounds has Frege for claiming that there *is* a thought that his words miss? If Frege's view is, in an important sense, unsayable, in what sense can it be a coherent view?

I suspect that it is Frege's viewing the problem in this way which explains his apparent lack of concern with the paradoxical nature of some of his remarks—his dismissing it as an awkwardness of language and his not introducing nonparadoxical locutions to replace them. For there is nothing to be gained by such a replacement unless the non-paradoxical locutions can really be used to provide an informative definition of concepthood. But concepthood cannot be defined, and Frege's sentences are merely elucidatory. Thus, their apparently paradoxical nature is of no concern unless it somehow prohibits their functioning as elucidations. Furthermore, the elucidatory purpose is not to fix the meaning of the term 'concept', but to get his readers to see how his Begriffsschrift notation expresses what, in the content associated with a sentence, is significant for purposes of inference.

Consider, for instance, Frege's discussion, in section 13 of *Basic Laws*, of the logical content of 'All square roots of 1 are fourth roots of 1'. He says, "Here we have the *subordination* of a concept under a concept, a *universal* affirmative proposition" (BLA 55). Of course this use of 'concept', like every other use of 'concept', makes Frege's words miss his thought. Since, when the term 'concept' is used correctly, we are forced to say that there are no concepts, there certainly can't be a concept that is subordinate to another concept. Or, to put it another way, to understand Frege's claim as the claim that the concept *square root of 1* is subordinate to the concept *fourth root of 1* is to understand it as a claim about two objects, the concept *square root of 1* and the concept *fourth root of 1*. On the other hand, while Frege's statement about the content of 'All square roots of 1 are fourth roots of 1' misses his thought, this is not to say that the purpose of Frege's statement has not been achieved. Frege's aim here is to get his readers to understand his logical notation—in particular the use of the quantifier and the conditional-stroke to express the content of a certain sort of claim. The remark that misses his thought here should not be regarded as a failed assertion that ought to have been stated in Frege's science of logic but, rather, as a hint that should make the nature of his logical regimentation clearer to his readers. There is no confusion or paradox involved in the use of this regimentation. If we do not assume that the term 'concept' is to be used in the statement of a central doctrine in a science of metaphysics or semantics, there is no reason to worry about the claims in section 13.

There might, however, be an objection. While it seems true that Frege has successfully used the term 'concept' to elucidate the translation of certain quantified sentences, elucidation is only appropriate when there can be no objective justification. Thus, it seems important to show that the claim about how 'All square roots of 1 are fourth roots of 1' is to be translated into Begriffsschrift notation is not a nonprimitive claim of some systematic science. In fact, it is clear that this claim about translation cannot be a nonprimitive truth of a systematic science. Frege's elucidation does not concern the complete translation of this sentence but, rather, the Begriffsschrift representation of a particular constituent(s) of the thought expressed by the sentence; that is, the Begriffsschrift representation of the constituent(s) that is expressed by 'All _____ s are _____ s'. For anyone who already understands what is expressed by this locution and who already understands the Begriffsschrift symbols, there is nothing more to say. Frege's symbols are simply a schematic representation of this content. In effect, Frege is giving an analysis of a long-established expression. And, as Frege says in "Logic in Mathematics," the agreement of such an analysis with the sense of a long-established sign "can only be recognized by immediate insight" (PW 210).[14]

The appropriateness of the representation can be supported by looking at how Frege's symbols are employed in inference and looking at the inferences in which everyday sentences making use of the everyday locution can figure. But this cannot constitute an objective argument, since it is clearly part of the presystematic setting up of a logically

---

[14]Actually, in the passage from "Logic in Mathematics" Frege is talking about the analysis of a simple sign, and one might suspect that 'All _____ s are _____ s' cannot be a simple sign. But this is not at all clear. Frege gives the following criteria, in *Basic Laws*, for recognizing a simple sign: "Any symbol or word can indeed be regarded as consisting of parts; but we do not deny its simplicity unless, given the general rules of grammar, or of the symbolism, the meaning of the whole would follow from the meaning of the parts [*aus den Bedeutungen der Theile die Bedeutung des Ganzen folgen würde*], and these parts occur also in other combinations and are treated as independent signs with a meaning of their own" (TWF 151/GGA vol. 2, 79). Now in the above expression, the parts that can occur in other combinations are 'All' and 'are'. But 'are' is not treated as an independent sign with a Bedeutung. In addition to expressions of the above type, it typically occurs either as a copula, as in

"On Sense and Meaning" and "On Concept and Object" are titles of articles written by Frege.

or as part of a quantifier-expression, as in:

There are even integers.

But 'are' does not have a Bedeutung of its own; nor does 'all'.

perfect language that can replace part of our everyday discourse. The translation of 'All _____ s are _____ s' is dependent on the appropriateness of the whole structure of Begriffsschrift.

It is the use of sentences that miss his thought which enables Frege to set up a logically perfect language to express inferences. And, if the appropriateness of his language was not evident to his contemporaries, it is evident to us today. Furthermore, given that Frege talks of elucidation as achieving mutual understanding by figurative expression and guesswork, there seems little reason to prohibit apparently paradoxical statements or statements that miss his thought. The only circumstance under which the use of an expression that misses his thought would count against Frege's view would be a circumstance in which such an expression is needed for a statement in his systematic science.

An interesting question remains. Frege talks about the simplicity of the notions of concept and object. Yet neither 'concept' nor 'object' is a primitive term of logic. In what sense are these notions simple? The success of Frege's communication of his notation will depend on his readers' sharing his presystematic views on which inferences are correct and understanding the sense in which his decompositions of sentences into their constituents mirror an analysis of logical constituents of the thoughts expressed, that is, constituents that play a role in inference or whose recognition (implicit or explicit) is involved in our recognition of the legitimacy of inferences. Frege says, in his explanation of the logical constituents of the thought expressed by '2 is a prime number', "The decomposition into a saturated and unsaturated part must be considered a logically primitive phenomenon which must simply be accepted and cannot be reduced to something simpler" (CP 281). If concepthood is characterized by predicative nature or unsaturatedness, our recognition that some expression stands for a concept is the result of a logically primitive activity. This logically primitive activity is involved in all inference.

Because the recognition of the notion 'concept', as well as of the notions 'function' and 'object', is involved in this logically primitive activity, Frege can appeal to his readers' recognition of these notions in the explanations of the differences between fundamental symbols of Begriffsschrift and how they can be combined. There are no symbols for objecthood and concepthood because these notions only provide constraints on the formulation of substantive expressions; there is no explicit role for them to play in the statement of laws of logic.

In many respects, the status of concepthood, functionhood, and objecthood is similar to the status of the notion of truth. Frege says that the term 'true' is indefinable: "Truth is obviously something so primitive and simple that it is not possible to reduce it to anything still simpler" (PW 128–129). Furthermore, the word 'true' seems defective in much the way 'concept' and 'function' are. Frege says, of the word 'true', that it "seems to make the impossible possible," and that this attempt "miscarries" (PW 252). And at least some uses of the word 'true' also seem empty in much the way uses of the word 'object' seem empty. Frege says, "It is also worth noticing that the sentence 'I smell the scent of violets' has just the same content as the sentence 'It is true that I smell the scent of violets'" (CP 354). He suggests that truth may not be a property in the ordinary sense. This certainly seems also to be true of concepthood and objecthood. Frege says, "The meaning of the word 'true' is spelled out in the laws of truth" (CP 352). One might, similarly, regard the meaning of the words 'function' and 'concept' as being spelled out by the logical or mathematical laws governing functions and concepts. Indeed, just as there is no Begriffsschrift term for the predication of concepthood or objecthood, there is no Begriffsschrift term for the predication of truth. Yet, just as an understanding of concepthood and objecthood is involved in our understanding of the rules that govern Begriffschrift notation, an understanding of truth is involved in another sort of rule governing Begriffsschrift, the rule of inference.

For all the difficulties with the word 'true', the notion of truth is, for Frege, a fundamental logical notion. Most of his writings about the nature of logic and most of his unpublished attempts at writing introductions to logic contain extensive uses of the words 'true' and 'truth'. These uses fall into two categories. One of Frege's uses is to describe logic as having truth as its aim. The other is in explaining his primitive laws and rule of inference. No defense of the use of a primitive logical law in logic, where our aim is truth, is necessary, because its truth is evident from its sense (CP 405). But a rule of inference cannot be expressed, as a logical law can be, by a formula of Begriffsschrift. A rule of inference is not something that is true or false but is, rather, a rule that licenses making a judgement on the basis of another. In order to explain why the use of his rule of inference is appropriate, given the aim of logic, Frege needs to explain why a judgement licensed via his rule of inference on the basis of a truth must, itself, be true. For such an explanation, the use of 'true' or some cognate of this term is indispensable.

His explanation of the legitimacy of his rule of inference, *modus ponens*, in *Basic Laws* is that if the consequent were not the True (were not true) then, since the antecedent is the True, the conditional would be the False. Once his readers recognize the legitimacy of *modus ponens*, however, this explanation can be abandoned. In a note headed "My basic logical Insights," Frege says:

> How is it then that this word 'true', though it seems devoid of content, cannot be dispensed with? Would it not be possible, at least in laying the foundations of logic, to avoid this word altogether, when it can only create confusion? That we cannot do so is due to the imperfection of language. If our language were logically more perfect, we would perhaps have no further need of logic, or we might read it off from the language. . . . Only after our logical work has been completed shall we possess a more perfect instrument. (PW 252)

The suggestion of this passage is that, ultimately, once his Begriffsschrift is understood and used, Frege will have no more use for the word 'true'. After all, in such a circumstance there would be no need to discuss the aim of logic or worry about the justification of the rule of inference. The words 'concept', 'object', and 'function' would likewise become superfluous in such a circumstance. Our understanding of these terms was only necessary in order to teach us the language of Begriffsschrift and the rules for its use and to communicate the general laws of logic and of functions. We come to understand these terms by elucidation only to use them in elucidations. This is, perhaps, why the defects of these terms never really seem to worry Frege.

What about the apparent necessity of the term 'true' for the expression of claims about translation and synonymy? It is important to note, first of all, that Frege did not advocate abandoning everyday language for everyday use, only for scientific use. Certainly none of our everyday claims about synonymy or translation need be abandoned. But do Frege's words seem to indicate a rejection of the possibility of an objective science of linguistics? This is also not at all clear. After all, when Frege suggests that the word 'true' might ultimately be abandoned, he is talking about the use of this word for the sort of project in which he is engaged—the laying of the foundations of logic. Frege does not address the issue of whether or not some precise use of the word 'true' can be found which will allow the formulation of a logically perfect language in which to conduct linguistic or psychological research. But it is also important to note that, insofar as claims about synonymy or translation can be understood as concerning logic in Frege's sense, they cannot play a role in a systematic science.

Frege talks about the circumstances in which different sentences express the same thought in two letters to Husserl, dated November 1 and December 9, 1906. He says, in the first of these letters,

> In logic, one must decide to regard equipollent propositions as differing only according to form. After the assertoric force with which they may have been uttered is subtracted, equipollent propositions have something in common in their content, and this is what I call the thought they express. (COR 67)

In the second letter he says that, if logical analysis is to be possible, there must be an objective criterion for recognizing a thought as the same. He then says:

> Now it seems to me that the only possible means of deciding whether proposition A expresses the same thought as proposition B is the following. . . . If *both* the assumption that the content of A is false and that of B true *and* the assumption that the content of A is true and that of B false lead to a logical contradiction, and if this can be established without knowing whether the content of A or B is true or false, and without requiring other than purely logical laws for this purpose, then nothing can belong to the content of A as far as it is capable of being judged true or false, which does not also belong to the content of B. (COR 70)

While Frege's characterization of the means of deciding whether two sentences express the same thought looks to be a metatheoretic characterization, it should be clear that the actual demonstration that two sentences express the same thought can be given by a Begriffsschrift proof. For, given the above criteria, if A and B express the same thought, then it must be possible to show that the assumption that the content of A is false and that of B true leads to a logical contradiction. But this should be so just in case the conditional with antecedent B and consequent A is a logical truth, hence, just in case the conditional is provable in Begriffsschrift. This suggests that, at least within a logically perfect language, the issue of identity of thoughts is nothing more than a question of demonstration, from the laws of logic, of a biconditional.[15]

---

[15]On the other hand, Frege's characterization of the senses of identities in "On Sense and Meaning" suggests something rather different. He suggests that an identity $a = b$ can contain actual knowledge if and only if the two signs, a and b, have different senses. But suppose that a and b are proper names of logic and that the identity $a = b$ can be proved

One might respond that the demonstration that two sentences express *different* thoughts must be metatheoretic. But Frege's reason for requiring an objective criterion for sameness of thought is that without it logical analysis is impossible. There is no reason to suppose that a criterion for difference of thoughts is required to make logical analysis possible. Frege's view might well be that disputes over the correctness of a logical analysis should no more play a role in research in a systematic science than disputes over the correctness of a definition. In the case of a definition, Frege requires that the manner of its introduction preclude raising the possibility that it might require justification. He might also require that any use of a logical analysis in a systematic science be preceded with Begriffsschrift proofs of the conditionals expressed by putting one of the original or analyzing sentences in the antecedent position of the conditional-stroke and the other in the consequent position.

Thus, it seems that insofar as Frege is concerned with synonymy his concern can be addressed without the use of any metatheoretic terms. But Frege's discussion of the criteria for two sentences expressing the same thought seems to presuppose that the sentences in question are sentences in a logically perfect language. What is the connection of Frege's discussion with the synonymy of sentences of everyday language? It seems that the issue of the expression of the same thought by

---

from logical laws. Then, using only primitive logical laws, it is possible to prove the equivalence of $a = a$ and $a = b$. It would then follow from Frege's remarks about identity of thoughts that, since $a = a$ contains no actual knowledge, neither does $a = b$. In other words, no logically provable identity can contain actual knowledge. But this surely does not seem to be the sort of conclusion Frege would advocate. In fact, Frege says something in "Logic in Mathematics" which almost appears to contradict it: "The sign '$(16 - 2)$' is a proper name of a number. '$(17 - 3)$' designates the same number, but '$(17 - 3)$' does not have the same sense as '$(16 - 2)$' " (PW 232). It may seem to follow that the sense of '$16 - 2 = 16 - 2$' must differ from that of '$16 - 2 = 17 - 3$'. But if the truths of arithmetic follow from the laws of logic, these sentences should express the same sense. In a letter to Paul Linke of 1919, Frege more directly contradicts this. He says, of the object names '$7 + 5$' and '$6·2$': "But even though these object names have the same meaning, they have a different sense, and this is why the proposition '$7 + 5$ is $7 + 5$', '$7 + 5$ is 12', and '$7 + 5$ is $6·2$' express different thoughts; for the sense of part of a proposition is part of the sense of the proposition, i.e., of the thought expressed in the proposition" (COR 98). Of course, "Logic in Mathematics" was written in 1914, and the letter from which the above quotation was taken was written later. By the time these were written Frege seems to have given up his conviction that the truths of arithmetic are logical truths. Hence it may be that neither of these passages contradicts the claims about sameness of thought from the letters to Husserl. On the other hand, there is clearly a tension between Frege's characterization of sense as cognitive content and his taking sameness of sense to have to do with logical content.

different everyday sentences will arise for Frege only when both sentences are to be translated into a logically perfect language. But, if so, should we be able to determine whether the translations are correct? The discussion of the logical analysis of everyday sentences in "Logic in Mathematics" suggests that this is not an issue. In general the senses of everyday sentences and sentences of a logically perfect language cannot be identical, because the senses of everyday sentences are imperfect. The translation must involve replacing the terms expressing imprecise presystematic constituents of the original thought with terms that have a precise sense. But once presystematic terms are replaced by precise terms, Frege says, "We must . . . explain that the sense in which this sign was used before the new system was constructed is no longer of any concern to us, that its sense is to be understood purely from the constructive definition that we have given" (PW 211). The issue of whether sentences of everyday science have been correctly translated into sentences of a systematic science is not an appropriate subject for a systematic science, on Frege's view.

From the point of view of the purposes of science, everyday language is merely an imperfect instrument that is to be used in the construction of a more perfect instrument for purposes of science. Frege says:

> I believe I can best make the relation of my ideography to ordinary language clear if I compare it to that which the microscope has to the eye. Because of the range of its possible uses and the versatility with which it can adapt to the most diverse circumstances, the eye is far superior to the microscope. Considered as an optical instrument, to be sure, it exhibits many imperfections, which ordinarily remain unnoticed only on account of its intimate connection with our mental life. But, as soon as scientific goals demand great sharpness of resolution, the eye proves to be insufficient. The microscope, on the other hand, is perfectly suited to precisely such goals, but that is just why it is useless for all others. (BEG 6)

One might respond that in the case of the replacement of the eye with the microscope, the latter is an instrument for observing the same physical objects that are observed with the eye but with higher resolution. Thus, Frege's analogy seems to require that the logically perfect language merely express more precisely the same thoughts that were formerly expressed by the everyday language. It should be as important not to switch thoughts when we begin to use the logically perfect language as it is that we not switch slides when we begin to use the microscope.

On the other hand, the guarantee that the slides have not been switched when we begin to use the microscope cannot be achieved by comparing what we observe before putting the slide under the microscope with what we observe after putting it under the microscope. If the microscope is sufficiently powerful, there may be no discernible similarity between the appearance of the slide to the naked eye and its appearance under the microscope. Similarly, it may be that the guarantee that the sentences of the logically perfect language express the same thoughts that are expressed imperfectly in the everyday language may not be obtainable by comparison of our understanding of the pre- and postsystematic sentences. There may seem to be little discernible similarity between the senses associated with the everyday sentence '$0 \neq 1$' and the elaborate Begriffsschrift expression that, given Frege's definitions, this sentence abbreviates. The justification of Frege's belief that, in his substitution of Begriffsschrift expressions for our everyday numerals he has not switched content, will ultimately lie in its applications. His systematic arithmetic must, for instance, allow us to prove that if there are three Fs, four Gs, and no Fs that are Gs, then there are seven things that are either Fs or Gs.

Thus, the evaluation of sameness of thoughts expressed by different sentences in a language that is not logically perfect is no part of Frege's project. As he says, in the earlier of the two letters to Husserl,

> It cannot be the task of logic to investigate language and determine what is contained in a linguistic expression. Someone who wants to learn logic from language is like an adult who wants to learn how to think from a child. . . . The main task of the logician is to free himself from language and to simplify it. (COR 67–68)

Frege's task is not to capture the thoughts expressed by sentences of the presystematic science of arithmetic but to introduce thoughts expressed in a logically perfect language which can replace the thoughts expressed by the presystematic sentences.

As I said earlier, none of this shows that Frege would, in principle, object to a scientific study of the sameness of what is expressed by sentences in everyday language. It shows only that Frege's concern is not with understanding how everyday language functions. His concern is with replacing everyday language with a more perfect instrument. Although we have to use the imperfect instrument of everyday language to construct a more perfect instrument, it in no way follows that the success of this construction requires a study of the imperfect instrument. But there is a tension here. After all, Frege says "One can hardly deny that mankind has a common store of thoughts which is transmitted from one generation to another" (CP 160; see also CP 185). This common store of thoughts has been transmitted via ordinary language. Thus, there must be objective (if, perhaps, imperfect) thoughts expressed by ordinary language, and the objectivity of these thoughts should guarantee the appropriateness of a scientific investigation of them. Since our only access to thoughts is via sentences that express them, if our aim is to investigate precise thoughts, we will need to express them—and for this we will need to use sentences of a logically perfect language. I suggested earlier that Frege might have regarded everyday sentences not as imperfectly expressing precise thoughts but as expressing imperfect thoughts. It may be that there is room for a science dealing with the sameness of imperfect thoughts expressed by everyday language but, if so, Frege's criterion of sameness of thought do not apply. In a very important sense, then, Frege is not a philosopher of language.

I want to turn, finally, to an examination of the consequences of Frege's understanding of objectivity for the evaluation of the status of his philosophical writings. In *Foundations* Frege says that what is objective is what is subject to laws, what can be conceived and judged, what is expressible in words (FA 35). This formulation of the notion of objectivity looks dated in view of the later distinction between sense and Bedeutung. For what is expressible in words must be a sense, while what is subject to laws (e.g., the number one) need not be. But there is a unifying thread in this characterization of objectivity which survives in his later characterizations of the distinction between what is objective and subjective in "On Sense and Meaning" and "Thoughts." This has to do with intersubjectivity and science.[16]

---

[16]In the discussion that follows I have ignored the fact that Frege seems to acknowledge, in "Thoughts," the existence of thoughts that are not intersubjective and cannot be expressed in language. He says: "Now everyone is presented to himself in a special and primitive way, in which he is presented to no-one else. So, when Dr. Lauben has the thought that he was wounded, he will probably be basing it on this primitive way in which he is presented to himself. And only Dr. Lauben himself can grasp thoughts specified in this way" (CP 359). Dr. Lauben's thought—which only he can grasp—cannot be intersubjective. Also, since Frege goes on to say that, for purposes of communication, the word 'I' is used in the sense of 'he who is speaking to you at this moment', it seems that Dr. Lauben's thought cannot be expressed in language. It is also unclear that there is a sense in which Dr. Lauben's thought is independent of him. This seems to contradict

In "On Sense and Meaning" Frege says, "The idea is subjective: one man's idea is not that of another" (CP 160) and "This constitutes and essential disinction between the idea and the sign's sense, which may be the common property of many people, and so is not a part or a mode of the individual mind" (CP 160). Frege's argument in "Thoughts" that thoughts are objective begins with the question, "Is it at all the same thought which first that man expresses and then this one?" (CP 360). He also asks if thoughts, although they are not perceivable, can "nevertheless, like a tree be presented to people as identical" (CP 360). After providing a characterization of ideas, he says, "I now return to the question: is a thought an idea? If other people can assent to the thought I express in the Pythagorean theorem just as I do, then it does not belong to the content of my consciousness" (CP 362). He also says, "Not everything is an idea. Thus I can also acknowledge thoughts as independent of me; other men can grasp them just as much as I; I can acknowledge a science in which many can be engaged in research" (CP 368). The objectivity of what is expressed in the Pythagorean theorem seems to be tied to its role in intersubjective science. The Pythagorean theorem is, as Frege says in *Foundations*, expressible in words; it can be conceived and judged.

Although this understanding of objectivity seems to assimilate the realm of the subjective to that of the ineffable, Frege includes discussions of the realm of the subjective both in *Foundations* and in "Thoughts." Most of his *Foundations* discussion of what is subjective appears in section 26, where he tries to get his readers to see that all intuitions, pure and empirical, have both subjective and objective sides. He discusses two sorts of statements: statements involving ascriptions of colors and statements involving truths of geometry. The objective contents of such statements are constituted by the thoughts expressed by precise claims in systematic sciences. But Frege also says, in his discussion of ascriptions of color, "The word 'white' ordinarily makes us think of a certain sensation, which is, of course, entirely subjective" (FA 36). He argues that when we say that snow is white, we are not talking about this sensation but about an objective quality of snow which it might have even if the sensation were absent. Some evidence of this is that it makes perfect sense to say, "It *appears* red at present, but *is*

---

much of what Frege says about thoughts, and I am not convinced that it can be reconciled with his other views. For discussions of this passage, see Dummett, IFP 118–128; E. H. Kluge, *The Metaphysics of Gottlob Frege*, 199–201; Burge, "Sinning against Frege."

white." Also, he argues, if in attributing colors to objects, we were talking about the character of our sensations, it would not be possible for colorblind people to be justified in making attributions of color. But a colorblind person who cannot distinguish red from green by sensation can nonetheless be justified in attributing these colors to certain objects. For instance, such a person might recognize one of these colors by making a physical experiment. The characters of our sensations of colors, then, are part of the realm of the subjective.

A similar discussion of colors and the character of sensations appears in "Thoughts:"

> When the word 'red' is meant not to state a property of things but to characterize sense-impressions belonging to my consciousness, it is only applicable within the realm of my consciousness. For it is impossible to compare my sense-impression with someone else's. For that, it would be necessary to bring together in one consciousness a sense-impression belonging to one consciousness and a sense-impression belonging to another consciousness. . . . In any case it is impossible for us men to compare other people's ideas with our own. (CP 361)

The intersubjective content of the word 'red' shows that redness is not a property of our private sensations.

But there is something very odd about Frege's characterization of sense-impressions as part of the subjective realm. What is the status of Frege's discussion of sense-impressions? On Frege's view, the possibility of communication via everyday sentences is guaranteed only by the possibility of their replacement by sentences of a systematic science. Frege's point about sense-impressions is that their character cannot be described. But if the sentences of his descriptions of sense-impressions could be replaced by sentences of a systematic science, there would have to be an objective formulation of the property of being a sense-impression. Hence, sense-impressions, as understood above, cannot be subjective. This is absurd. It follows, then, that these discussions of the realm of the subjective cannot be understood as vague everyday expressions of objective claims, that they cannot be replaced by objective arguments in a systematic science. How, then, are they to be understood?

We can regard Frege's aim, in his descriptions of sense-impressions, not as describing subjective objects but as asking his readers to call up subjective mental images and sensations and recognize them as such. If so, Frege cannot be giving an objective description of the realm of the subjective, he can only be making suggestive remarks that will get his readers, if they are cooperative, to understand the significance of his notion of linguistic expressibility and systematic science. The nature of this activity comes out more clearly in his *Foundations* discussion of the subjective side of spatial intuition. Instead of asking his readers to imagine the mental images of a colorblind person, Frege asks his readers to imagine two beings who intuit only projective properties and relations, one of whom intuits as a plane what the other intuits as a point, and so on. What one will intuit as a line joining two points, then, would be intuited by the other as a line of intersection of two planes.

Now it is important to note that the the two beings intuit projective, not Euclidean, properties and relations. On Frege's view, we are not such beings. Just as non-colorblind people have no way of imagining sensory images that have the same character as a colorblind person's sense-impressions of red and green, we have no way of picturing the projective space intuited by these two beings. Furthermore, since in projective geometry what is true of planes is precisely what is true of points, these two beings would agree on all axioms and theorems of geometry.[17] And since these axioms and theorems are precisely what is objective about the pure intuitions of these beings, there is no objective difference between their pure intuitions. But Frege's story seems to indicate that there is a difference. He says:

> Over all geometrical theorems they would be in complete agreement, only interpreting the words differently in terms of their respective intuitions. With the word "point", for example, one would connect one intuition and the other another. (FA 36)

This difference in intuitions is a subjective difference.

It is not open to Frege to regard the above story, or anything else he can say about the intuitions of these beings, as a description of a subjective difference. How, then, is this example supposed to work? Frege seems to be assuming that these beings have intuitions of points and planes which differ, for them, in much the way our intuitions of points and planes differ. We cannot picture the images that 'point' and 'plane' call up in the projective beings and observe their difference, but we can picture the images these words call up in us and consider their

[17]A projective plane can be constructed from a Euclidean plane by adding, for each line in the original plane, exactly one point at infinity. Lines that were parallel in the original plane share a point at infinity in the projective plane. Furthermore, the points at infinity in this plane determine a single line at infinity. Two planes that are parallel in Euclidean space share a line at infinity when the points at infinity are added to each.

difference. Suppose we imagine the interchanging of 'point' and 'plane' by imagining two different pictures, both of which represent how a line is determined by two points (illustrations A and B). Of course there are objective differences here. These pictures are not mental images, they are physical drawings, and there are physical differences between them. However, these physical differences cannot be differences between mental images, for mental images are not physical objects. Also, if points are represented as in the picture on the right, we can picture two points that do not determine a line (illustration C). There is another objective difference. Any two points represented by the sort of figures in illustration A will determine a line. But Frege is assuming that his readers will feel that there is a difference between the mental pictures which is not expressed by this fact. The assumption is that there is a difference here which is comparable to the difference between the images of the two projective beings and, since that difference cannot be objectively expressed, there must be some difference between our two images which cannot be objectively expressed. Whatever this difference is must be subjective. If his readers recognize this difference, they will thereby recognize what Frege means when he talks about what is subjective. Frege will only be successful in this enterprise if he is granted his grain of salt, for his readers could easily claim to recognize

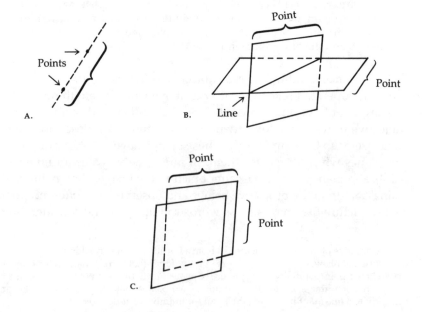

no difference other than the objective difference. To such readers, Frege has no response.

It is important to note that this reliance on a grain of salt is very serious. It is not just that Frege cannot give an objective argument that there is a difference between the mental images his story calls up in his readers. There is no real sense to be made of the claim that there is a difference or that his sympathetic readers will find a difference because, of course, there *is* no objective difference. Talk of the subjective realm or elements of the subjective realm can only be elucidatory. There is nothing to say about the elements of the subjective realm. What is subjective cannot be described. It follows, of course, that there is no way to make explicit a distinction between what is subjective and what is objective. Frege's subjective/objective distinction is no better off than his concept/object distinction. The role of Frege's discussion of objectivity can only be elucidatory—it can be of help in illuminating Frege's demands on science and persuading his reader that the demands are reasonable.

Since Frege says that mental images are subjective, he must regard any informative or useful talk about mental images as merely elucidatory. One might infer that this is meant as a critique of introspectionist psychology. But the situation is actually somewhat more complicated than it appears. It is, of course, obvious that on Frege's view introspectionist psychology is defective. For it is not a systematic science. But this defect is not peculiar to introspectionist psychology. On Frege's view, even arithmetic was defective before his work. And although Frege's work was to have eliminated the foundational defects of arithmetic, these defects remain in all other sciences. Chemistry, in its current state, is every bit as nonsystematic as introspectionist psychology.

It may seem that the difference between introspectionist psychology and chemistry is that, while chemistry is about physical (hence objective) objects, introspectionist psychology is about mental images (which are not objective). But this distinction cannot be made out. For it makes no sense to talk of what a science is about until proper foundations have been given. Frege seems to believe that before his foundations were given, nobody understood what arithmetic was about. This is not to say that anyone would disagree with the claim that arithmetic is about numbers but, rather, that nobody really understood what this claim meant. For, as Frege's quotations of mathematicians and philosophers show, to say that arithmetic is about numbers is not to deny that arithmetic is about ideas or shapes of inscriptions. In order to know what arithmetic is about, we must understand its fundamental concepts, including that of number. Frege discusses this situation at some length in "Logic in Mathematics" (PW 215–222), where he goes so far as to suggest that Weierstrass was unclear about the basic concepts of arithmetic and to ask "But how, it may be asked, can a man do effective work in a science when he is completely unclear about one of its basic concepts?" (PW 222). Frege's foundations are meant to show us what arithmetic is about. The fact that many mathematicians would disagree with him is irrelevant. Thus, the fact that people believe that introspectionist psychology is about inner mental images shows very little.

It is by no means obvious that it would be any more difficult to find a systematic foundation, in Frege's sense, for introspectionist psychology than it would be to find a foundation for any other science. The upshot of providing such a foundation for introspectionist psychology would be very similar to the upshot of Frege's foundation for arithmetic. First, although Frege's foundation for arithmetic requires some modifications in mathematical practice (more rigor, in particular), for the most part Frege's foundation legitimates the notion of proof as mathematical justification and allows mathematical work to go on more or less as before. Frege's foundations are not meant to transform mathematics into something unrecognizable. Similarly, were such a foundation provided for introspectionist psychology, it would have to legitimate enough of the standards actually in use in such a way as to allow work in this field to go on more or less as before. Second, a systematic foundation for introspectionist psychology would show, as Frege's foundation for arithmetic does, what the subject matter of this science is. Frege understands his foundation for arithmetic to show that numbers are properties of concepts. It makes no difference that many mathematicians had claimed that numbers were inscriptions or even that so distinguished a mathematician as Weierstrass appears to disagree with Frege's conception of numbers.

Similarly, a foundation for introspectionist psychology would, on Frege's view, show that its subject matter is not really mental images. To say this is not to say that the case is so different from that of numbers. After all, systematic arithmetic is not really about what counts as numbers on some mathematician's presystematic notion of number. Insofar as mathematicians regard numbers as inscriptions, the numbers of Frege's systematic science will be different from the presystematic notion of numbers. Anyone who is looking for a systematic science of inscriptions will not find it in Frege's systematic arithmetic. And insofar as mental images are regarded as the sort of subjective entity Frege is attempting to hint at, there will be no systematic science that deals with mental images. None of this need cast suspicion on the conclusions drawn by people who thought psychology was about mental images, any more than Weierstrass's unclear understanding of the nature of the numbers need cast suspicion on his proofs. Thus, in an important sense, Frege is actually not criticizing introspectionist psychology.[18] The upshot of a successful attempt at providing a foundation for introspectionist psychology would be that, in what looks to be talk about mental images, there are actually objective psychological claims.

This discussion should highlight the ineffability of the subjective. Although there is no compelling reason to suppose that foundations cannot be found for introspectionist psychology, the upshot of finding

[18]There is, of course, an entirely different sense in which Frege might be taken to have objections to introspectionist psychology. For he does talk about what he regards as objective psychology. Frege's mention of psychological laws usually occurs in an attempt to contrast these laws with logical laws. Both sorts of laws concern acts of judging. The logical laws are normative: one *ought* to judge according to these laws if one wants to judge truly. On the other hand, the psychological laws are more like laws of human nature. They are generalizations about the conditions under which we will in fact make particular judgements. Logical laws tell us what reasons actually justify making certain judgements, while psychological laws tell us what processes or reasons usually lead people to make these judgements. Frege says: "The grounds on which we make a judgement may justify our recognizing it as true; they may, however, merely give rise to our making a judgement, or make up our minds for us, without containing a justification for our judgement. Although each judgement we make is causally conditioned, it is nevertheless not the case that all these causes are grounds that afford a justification. . . . The causes which merely give rise to acts of judgement do so in accordance with psychological laws" (PW 2).

There is a good deal of current research in psychology which, presumably, Frege would endorse. In the introduction to *Judgment under Uncertainty: Heuristics and Biases*, edited by Daniel Kahneman, Paul Slovic, and Amos Tversky, Kahneman and Tversky outline a list of heuristics in accordance with which people actually judge. One such heuristic is that "a sequence of events generated by a random process will represent the essential characteristics of that process even when the sequence is short" (7). Thus, if a coin has been tossed ten times and, each time, has come up heads, people will judge that it is more likely to come up tails on the next toss. This is clearly a law according to which people do judge rather than one according to which people ought to judge, for the probability of the coin coming up tails on the next toss is .5.

It is clear why Frege finds this sort of psychology of more interest than introspectionist psychology. For such work is much more likely to affect people's understanding of his own work on the true laws of thought. For instance, any demonstration that people do not judge in accord with Frege's laws of thought is evidence for the normative value of setting out the laws of thought. A study of the incorrect inferences people actually make might also be useful for finding strategies for communicating the correct rules for inference. However, it is important to note that Frege's endorsement of this sort of psychology is a consequence of his interests, not of his views. This sort of psychology, due to its lack of systematic foundations, is as suspect as introspectionist psychology.

such a foundation would be that the subject matter of this science is not what it has generally been taken to be. If claims that seem to be about our mental images are objective, they are not actually claims about mental images. But if they are subjective, they have no nonelucidatory substance. And any elucidatory substance such claims might have only amounts to getting us to understand objective truths. Thus, there is nothing to say about mental images. Anything I may have appeared to say about mental images must have missed my thought.

What is the upshot for the interpretation of the significance of Frege's philosophical writings? I argued earlier that elucidation has a substantive role to play in Frege's scientific project of introducing a logically perfect language and logical laws and proving the truths of arithmetic in this language using these laws. I also argued earlier that virtually all of Frege's philosophical writings can be viewed as playing an elucidatory role for this project. But these arguments are not sufficient to support a reading on which Frege viewed all his philosophical writings in this way. There is, however, some evidence that he might have done so (or might have been pushed to do so). At least one of these writings, "On Concept and Object," is defective unless it is viewed as playing an elucidatory role. And, while Frege explicitly recognizes and discusses this defect, he is unwilling to take it seriously. Frege's attitude toward this defect as well as toward the importance of a successful elucidation of his use of 'function' and 'concept' in setting up the notation—particularly given his aim of convincing logicians to adopt his notation in favor of more traditional notations—suggests that the role of "On Concept and Object" is, on Frege's view, elucidatory.

The significance of the defect in Frege's use of 'concept' cannot be limited to the elucidatory passages of *Begriffsschrift* and *Basic Laws* in which he sets up his notation and the one article in which he discusses his notion of concept. For Frege's understanding of concepthood and his use of the term 'concept' also plays an important role in his discussion of arithmetic. The illustration of the awkwardness of language can be re-created through an examination of his claim that assertions of number are about concepts. The evidence that suggests that "On Concept and Object" must be viewed as playing an elucidatory role suggests, also, that one of the central claims of *Foundations* must be viewed as playing an elucidatory role. I also argued that, if all claims for whose statement the use of the term 'concept' is required must miss Frege's thought, then so must all claims for whose statement the use of the term 'object' is required. Thus, another of the central claims of *Foundations*, the claim that numbers are objects, must also miss his thought and can play only an elucidatory role. Another major theme of *Foundations* is the objectivity of the truths of arithmetic. But I have argued, above, that an examination of Frege's discussions of what is subjective reveals the same sort of awkwardness found in his discussions of concepthood. And by the same sort of argument about the consequences of the problems with the word 'concept' for Frege's use of the word 'object', the awkwardness of Frege's use of 'subjective' infects his use of 'objective'.

All three of the central claims of *Foundations* mentioned above—that assertions of number are about concepts, that numbers are objects, that the truths of arithmetic are objective—can, it seems, be shown (via the proofs of *Basic Laws*) but not stated. Furthermore, although I have argued that the motivation of Frege's overall project can be attributed to a substantive philosophical perspective, the defects in the explicit statements at least of the first two of these claims do not come from the details of this perspective but, rather, from consequences of some of Frege's most basic and explicitly stated views about logic. Similar arguments can be given for most of the views expressed in Frege's philosophical writings. Most of these writings are concerned with how Frege's logical notation is to be understood, that is, with features that are not expressible in it.

It might seem to be a consequence of this view that, ultimately, once the project of *Basic Laws* has been completed, Frege's philosophical work should be dismissed. But this is not at all obvious. It is important to remember that Frege's mathematical project is not motivated by the demands of science or mathematics. Indeed, to reform mathematics as Frege suggests would clearly impede mathematical research. Frege's elucidations and his mathematical work, his call to find systematic sciences whose logically perfect thoughts can replace those of everyday science, are responses not to the demands of mathematics but to the demands of an epistemological view loosely derived from Kant. To see what status Frege's philosophical work can have, it may help to consider what role Frege means it to play for mathematicians. Two obvious possibilities are that it is meant to play the role commonly played by the introductory remarks in mathematics texts and that it is meant to provide a guarantee of the legitimacy of mathematical research. In either of these cases Frege's philosophical work can be viewed as elucidatory and it would be possible to abandon the elucidation once its aim is achieved. But neither seems accurately to describe the purpose of Frege's philosophical writings.

Although Frege's writings contain extensive criticisms of introductory sections of mathematics texts, the point of these criticisms is not so much that the introductory sections are inadequate as introductions for students but that the comments by mathematicians exhibit an inadequate understanding of the numbers. Frege himself suggests that young students do not have the intellectual maturity to understand the fully rigorous mathematics he advocates (PW 221). Thus, his objections to these introductory sections of texts are not likely to have anything to do with their pedagogic efficacy. Mathematicians do not, as a rule, write discursive papers on the nature of number. Thus, it is likely that Frege's criticisms of the views of mathematicians are based on the motivational asides in articles and introductory sections of texts because this is the only explicit written evidence about how mathematicians think about what numbers are.

It also seems unlikely that Frege's aim is to provide a legitimation for arithmetic, which, given his epistemological views, is necessary if mathematicians are to continue with their work in good conscience. Although Frege's criticism of Hilbert might be taken to suggest that on his view mathematical work generally requires legitimation, this is clearly wrong. Frege did not regard the demonstration that arithmetic can be replaced by a systematic science as something that is required if we are to distinguish arithmetic from illegitimate enterprises. At no point in his career did he ever seriously consider the possibility that arithmetic might be illegitame or seriously suggest that any of the mathematical results concerning the numbers are suspect. Even the failure of his *Basic Laws* led him, not to doubt the results of arithmetic, but, rather, to suspect that his particular strategy for attempting to systematize arithmetic was wrong. Frege's elucidations are not meant to be instrumental in the legitimation of the work of mathematicians.

All this suggests that Frege's philosophical work is not meant to serve the pragmatic aims of mathematicians at all. Although he advocates some reform of mathematical practice, his replacement of arithmetic with a systematic science is not meant to influence mathematicians to replace their actual proofs with gapless proofs in a logically perfect language. Such proofs would be far too complex to be useful. Frege says:

We simply do not have the mental capacity to hold before our minds a very complex logical structure so that it is equally clear to us in every detail. For instance, what man, when he uses the word 'integral' in a proof, ever has clearly before him everything which appertains to the sense of this word! And yet we can still draw correct inferences, even though in doing so there is always a part of the sense in penumbra. (PW 222)

It is, of course, conceivable that, on some occasions, Frege's Begriffs-schrift might allow mathematicians to pinpoint confusions with greater accuracy. For his notation makes the expression of great complexity possible. However, even if his standards for rigor in proofs and defini-tions were accepted as correct by working mathematicians, proofs and definitions satisfying these standards could only rarely play a role in actual mathematical practice. In general, it would be psychologically impossible for mathematicians to work with proofs and definitions that meet Frege's criteria for rigor. Frege acknowledges this in the preface to *Basic Laws*. He comments that Dedekind, in *Was sind und was sollen die Zahlen?* gets more results from the laws of arithmetic than Frege will in *Basic Laws*. This is a result of the different standards of rigor applied in these two works. But, Frege adds,

> I do not say this as a reproach, for his procedure may have been the most appropriate for his purpose; I say it only to set my intention in a clearer light by contrast. . . . Generally people are satisfied if every step in the proof is evidently correct, and this is permissible if one merely wishes to be persuaded that the proposition to be proved is true. (BLA 4)

Frege's own aim is to do more than to prove that his propositions are true, his aim is also to gain insight "into the nature of this 'being evident'" (BLA 4). This, he goes on to say, is not generally the aim of mathematicians, but he does not argue that it should always be the aim of mathematicians. The complex definitions and rigorous proofs that constitute Frege's actual foundations for arithmetic cannot be held in the mind of every mathematician who uses the integers. Thus, the details of Frege's foundations will, in practice, end up in penumbra.

What effect, then, does Frege want his foundations for arithmetic to have on mathematicians? Frege does want mathematicians to eliminate some of the remarks found in introductory sections of mathematics texts and articles which he regards as silliness, but these remarks have little to do with the actual research. He also wants to prohibit the use of certain sorts of definitions which do not meet his strict criteria, espe-cially definitions of functions which only determine values on a limited range of arguments. Aside from these changes, his work does not seem designed to have much effect on mathematical practice. It seems unlikely that Frege's extensive philosophical writings could have been written only with the aim of effecting these somewhat modest changes. Certainly his arguments for his restrictions on definitions, his greatest mathematical departure from traditional mathematics, are independent of most of his philosophical arguments.

It seems that one of his goals is to force mathematicians to regard mathematics differently. For the interest in the proofs of *Basic Laws*, a book of proofs, lies not in the establishment of heretofore unproved truths, nor in the mathematical novelty of the constructions. Rather, the interest in these proofs lies in their connection with a view of numbers which does not have any practical mathematical consequences. Frege's view seems to be that, if the work of a mathematician is to be intellectually mature, the mathematician must know what it is that she or he proves (PW 216). And in order for a mathematician to meet Frege's standards for such knowledge, she or he must become more philosophically sophisticated or, as Frege imagines a hostile mathematician putting it, be "infected by philosophy" (PW 216). Frege says, "A mathematician with no element of philosophy in him is only half a mathematician" (PW 273). The aim is to suggest not that philosophy is a part of science or mathematics but, rather, that there is an emptiness to the intellectual enterprise of giving purely abstract mathematical proofs.

This brings us back to the question of whether, in saying that these views do not have the objective status of truths of science, I am denigrating Frege's achievement. I do not think I am. For I see no reason to think that the only valuable intellectual activity is the establishment of objective scientific truths. While a version of Frege's logical notation has had lasting impact on mathematics and science, this is not our only reason for studying his writings. Even if my conclusions about Frege's notion of objectivity are wrong, and his purpose was primarily the establishment of philosophical truths, the value of his work does not lie in his having established philosophical truths. Although his views are largely recognized as failed views, this recognition does not prevent us from finding illumination in them. Some find in Frege's work a motivation to mathematize philosophy. I find something rather different. Frege says,

> After refuting errors, it may be useful to trace the sources from which they have flowed. One source, I think, in this case is the desire to give definitions of the concepts one means to employ. It is certainly praiseworthy to try to make clear to oneself as far as possible the sense one associates with

a word. But here we must not forget that not everything can be defined. If we insist at any price on defining what is essentially indefinable, we readily fasten upon inessential accessories and thus start the inquiry on a wrong track at the very outset. (CP 381)

At least one of the morals that can drawn from Frege's work, no matter what its flaws, is a moral about the limitations of mathematization which is an illustration of where attempts at precision break down.

APPENDIX A

The True

ALTHOUGH the Begriffsschrift notation introduced in *Basic Laws* contains no primitive signs that are proper names, it may seem, nonetheless, that there is a hidden primitive proper name: "the True" (*das Wahre*). For in section 31, which is titled "Our Simple Names Denote Something [*etwas bedeuten*]," BLA 87/GGA xxviii,[1] Frege says, "We start from the fact that the names of truth-values denote something [*eine Bedeutung haben*], namely, either the True or the False" (BLA 87/ GGA 48). But there is a puzzle with this claim which must be cleared up before we can address the issue of whether Frege has a hidden primitive proper name. Has Frege, at this point in *Basic Laws*, shown that there are any names that are names of truth-values? Any name for a truth-value must be constructed from the names of Begriffsschrift, and the purpose of this section is to argue that these names have Bedeutung. If one of these names does not have Bedeutung, then a complex expression in which it appears and which looks to be the name of a truth-value will not have Bedeutung. It seems that the assumption that there are names of truth-values presupposes that at least some simple names involved have Bedeutung.

If Frege's discussion in section 31 is to make any sense, it is important to consider the significance of one of the differences between Frege's

[1]Note that by 'names' (*Namen*), Frege does not mean 'proper names', since the simple names he discusses are all function-names.

logic and Aristotelian or Boolean logic. This difference is that Frege starts with judgements rather than with concepts (see CN 94, PW 17, PW 253, COR 101). He claims that concepts are arrived at by splitting up a content of possible judgement or, in Frege's later terminology, a thought. He says, for example,

> Also, without symbols we would scarcely lift ourselves to conceptual thinking. Thus, in applying the same symbol to different but similar things, we actually no longer symbolize the individual thing, but rather what [the similars] have in common: the concept. This concept is first gained by symbolizing it. (CN 84)

The unsaturated nature of concepts prevents us from symbolizing a concept on its own. When Frege talks about symbolizing the concept, about applying the same symbol to different things, he is talking about a symbolization that occurs in the context of articulating and making judgements about these different things. This is made clearer in an unpublished paper comparing Begriffsschrift with Boole's calculus, where Frege says:

> And so instead of putting a judgement together out of an individual as subject and an already previously formed concept as predicate, we do the opposite and arrive at a concept by splitting up the content of possible judgement. Of course, if the expression of the content of possible judgement is to be analysable in this way, it must already be itself articulated. We may infer from this that at least the properties and relations which are not further analysable must have their own simple designations. But it doesn't follow from this that the ideas of these properties and relations are formed apart from objects: on the contrary they arise simultaneously with the first judgement in which they are ascribed to things. (PW 17; see also COR 101)

This is not to say that thoughts are invariably grasped before their unsaturated constituents. He suggests (CP 390, COR 79) that a person can grasp an entirely new thought via a sentence constructed out of words she or he has seen before which express the constituents of the new thought. But this situation can occur only if the person in question already understands what is expressed by the constituents of the sentence and that understanding comes from splitting up judgements. Thus, if we have a sentence that expresses a true thought we can, through analysis of the content of that thought, arrive at the senses of the sentence's constituent parts and, via these senses, the parts' Bedeutung.

Thus, the success of Frege's introduction of primitive Begriffsschrift symbols that have Bedeutung depends only on their being introduced in the context of a representation of a thought with a truth-value. But how do we defend the claim that there are expressions that express true (or false) thoughts? This claim is really one of Frege's fundamental assumptions. To see why, it suffices to put together some of his comments from "On Sense and Meaning" and "On Concept and Object." In both articles he claims that mankind has a "common store of thoughts which is transmitted from one generation to another" (CP 160; see also CP 185, CP 371). He also argues, in "On Sense and Meaning" (CP 163), that it is impossible for anyone, even the sceptic, to refrain from judging altogether. Indeed, it is impossible to argue that no thought we have expressed is either true or false. For such an argument must itself depend on a judgement that a particular claim is true. Furthermore, the view that we do express truths does not seem to be one that Frege would regard as controversial. Even if most of our ordinary language is hopelessly vague, to deny that "Every object is identical with itself" expresses a truth would be, on Frege's view, to exhibit a hitherto unknown form of madness (BLA 14). Finally, the purpose of Frege's extended introductions of his Begriffsschrift, both in his original publication and in *Basic Laws*, is to convince us that this notation does, indeed, allow us to express the content of logical truths (e.g., "Every object is identical with itself").

The assumption in section 31 that there are expressions of true thoughts is not, in itself, a problem. Of course, the assumption with which he starts is somewhat stronger than this. He also assumes that expressions that express true thoughts name truth-values. This, too, is one of the views for which he argues in "On Sense and Meaning." In his argument that even the sceptic must believe that we express some true thoughts, he says, of the True and the False, "These two objects are recognized, if only implicitly, by everybody who judges something to be true—and so even by a sceptic" (CP 163). Frege says little else about what these objects are, except that they are the circumstance that a sentence is true or false. This may look to be an indication that the True and the False are logically simple and what they are can be indicated only by hints, but I do not think this is obvious. It may also seem that there is something especially mysterious about the notion of truth-value. But this, I think, is clearly mistaken. I have argued that Frege's notion of having Bedeutung, or denoting something, has to do with precision of expression. Thus, a sentence all of whose terms have Bedeutung must itself have Bedeutung. Furthermore, since sentences do not express something unsaturated, they must be object-expressions. But to say that sentences name objects is not to say that we can recognize some object, by some sort of ostension apart from language or by some sort of linguistic description that is not a sentence, and identify that object as the thing named by a particular sentence.

But, if the explicit statement of his assumption is not problematic, there seems to be a hidden problem. Frege goes on to argue that some of his primitive first-level function expressions denote something when they are supplemented by a name of a truth-value. How do we know that any Begriffsschrift expressions are names of truth-values? The argument for this seems to appear in section 10. Frege says that, without violating Basic Law V,

> It is always possible to stipulate that an arbitrary course-of-values is to be the True and another the False. Accordingly let us lay it down that $\grave{\varepsilon}(\text{———}\varepsilon)$ is to be the True and $\grave{\varepsilon}(\varepsilon = (\text{⊤}\varnothing\frown a = a))$ is to be the False. (BLA 48)

This is somewhat mysterious, especially in light of what I have said above about our not needing any sort of ostension or nonsentential expression by which to recognize truth-values. For these stipulations appear to give us nonsentential names for the True and the False. On the other hand, we might view these stipulations as assigning a logically simple Bedeutung, the True or the False, to each. But there is also something odd about this circumstance. Consider how this is to be used in Frege's argument, in section 31, that his primitive expressions have Bedeutung. For the sake of simplicity, I will concentrate primarily on the True, but all remarks about the status of the stipulation concerning the True below apply equally to the stipulation concerning the False.

The central assumption in section 31 is that there are, in Begriffsschrift, names of the True and the False. This should be a consequence of the stipulation in section 10. But what, exactly, is stipulated in section 10? Since the expressions 'the True' and 'the False' do not appear in Begriffsschrift notation, and since stipulations of this sort are generally regarded as the assignment of a sense and Bedeutung to an expression, it does not seem unreasonable to assume that Frege's aim is to assign a sense and Bedeutung to the Begriffsschrift expressions: $\grave{\varepsilon}(\text{———}\varepsilon)$, $\grave{\varepsilon}(\varepsilon = (\text{⊤}\varnothing\frown a = a))$. Yet the primitive expressions that, Frege argues in section 31, have Bedeutung are constituents of the expressions that, he stipulates in section 10, stand for the True and the False. It seems, then, that in section 10 Frege is violating either his principles of definition or his principles of introducing a sign for something logically simple. For the introduction of a sign, by definition or elucidation, should be the introduction of a simple sign, a sign whose constituents do not play logical roles. And the signs that, by Frege's stipulation, are to stand for the True and the False have constituents that play logical roles independent of these expressions.

One might suspect that the point is to determine values for a borderline case for the course-of-values function, the case when the first-level function to which it is applied is the horizontal function. But this seems wrong. For if we do know what the True is, then, given Frege's explanation of the horizontal function, there should be no more mystery about its course-of-values than about the course-of-values of any other function.

Frege's talk about our attaining concepts by decomposing thoughts suggests another response. If the symbols that, by Frege's stipulation, stand for the True and the False express thoughts, we might be able to gain the concepts involved (hence the Bedeutung of some of the constituents of the expressions) by decomposing the thoughts. The problem with this response is that the expressions in question do not express thoughts. The upshot seems to be that Frege has made a stipulation that violates his principles and that the argument in section 31 is circular.

I do not think this is truly an upshot of what Frege has done. In order to interpret these sections, we will need to look more closely at the aim of section 10. I began by interpreting Frege's stipulation as the assignment of a logically simple object, either the True or the False, to each of two complex expressions of Begriffsschrift. On this interpretation Frege is making at least some serious errors. However, I think that there is another interpretation of section 10 which is much more plausible than the interpretation given above. As I mentioned above, Frege says little in "On Sense and Meaning" about what the True and the False are, and it is tempting to infer that these are logically simple objects. But there is another way of looking at Frege's claim that sentences designate truth-values.

To say that truth-values are what sentences designate is no more mysterious than to say that numbers are what numerals designate. In each case the description fails to be a precise specification—in each case we are given no means for deciding, for instance, whether or not Frege's pocket watch is one of the objects in question (see CP 303)—but it does not follow that the nature of numbers or truth-values is entirely mysterious or logically simple. Just as what we know about numbers is given by what we regard as the laws of arithmetic and their applications, what we know about truth-values is given by what we regard as the logical laws (the laws of truth) and their applications. Just as we are entitled to pick out, for the numerals, any objects that will allow us to replace arithmetic with a systematic science, we are entitled to pick out, for sentences, any objects that will allow us to introduce a systematic science of logic. The upshot, then, is that Frege's introduction, in "On Sense and Meaning," of the True and the False does not answer our questions about what these objects are, any more than *Foundations* answers our questions about what the numbers are. The real answer to both questions can come only via the exhibition of an appropriate systematic science.

But if we do not yet know, in section 10, what the True and the False are, and if 'the True' is not a Begriffsschrift expression, what is being stipulated in section 10? I have, thus far, avoided discussion of courses-of-values or extensions of concepts. Given the apparently technical nature of these notions, one might suppose that they are to be defined. In fact, however, Frege says no more about what courses-of-values of functions are than he says about what truth-values are. He says that he assumes his readers know what the extension of concepts is (FA 80), and he introduces the course-of-values (or value-range) of a function (in "Function and Concept") via an informal statement of Basic Law V. He says, of the equation $x^2 - 4x = x(x - 4)$:

> And if we so understand this equation that it is to hold whatever argument may be substituted for x, then we have thus expressed that an equality holds generally. But we can also say: 'the value-range of the function $x(x - 4)$ is equal to that of the function $x^2 - 4x$', and here we have an equality between value-ranges. (CP 142)

He then says that this is indemonstrable and "must be taken to be a fundamental law of logic." In a footnote to section 147 of volume 2 of *Basic Laws* he says that "mathematicians have already made use of the possibility of our transformation" (TWF 160).

Let us suppose that Frege assumes we know in section 10, not what the True is but, rather, what the course-of-values of the horizontal is and what the course-of-values of the concept *identical to the denial that every object is self identical* is. That is, let us suppose Frege assumes that, given his elucidations of his primitive Begriffsschrift expressions, we know what the expressions '$\grave{\varepsilon}(\varepsilon = (\frown\!\smile\!\frown a = a))$' and '$\grave{\varepsilon}(\!\!-\!\!-\!\!\varepsilon)$' stand for. Since 'the True' is not a Begriffsschrift expression, he cannot be stipulating the sense and Bedeutung of this sign. What, then, is he stipulating?

It is important to remember that the success of Frege's introductions of his primitive signs depends on our understanding the signs as expressing notions that are already operating in our understanding of the logical structure of our everyday expressions. The introduction of the conditional-stroke, for instance, depends on our antecedent recognition of how what is expressed by certain expressions (conditionals) can be used in inferences. Once Frege has introduced some primitive Begriffsschrift terms out of which a sentential expression can be constructed, say, the conditional-stroke and the quantifier, there is no difficulty for us in understanding what is expressed by a sentence constructed out of those signs. There is no difficulty in our understanding the sense of

and, indeed, there is no difficulty in seeing that the thought expressed is true. However, given that this expression is not unsaturated, we need more. We need to know which identity claims in which this expression occurs are true and which false. There is no difficulty if the identity claim is expressed by an identity sign flanked by two sentential expressions. In this case the identity is true if either both sentences are true or both are false, and it is false otherwise. But if on one side of the identity sign there is a sentential expression and on the other side there is another sort of expression, none of Frege's remarks about truth-values will suffice, along with information about the truth or falsity of the sentence, to enable us to evaluate the truth or falsity of the identity. This is the issue that section 10 is supposed to settle. The stipulation is not of a sense and Bedeutung for the expression 'the True', nor is it a stipulation of the borderline cases for the horizontal and combination of Begriffsschrift symbols which appears in the expression correlated with 'the False'. Rather, what is stipulated is the value of sentential expressions that express truths or falsehoods.

What makes this stipulation legitimate is that it does not violate the law about the second-level course-of-values function or any of the laws of truth. It is immaterial that, on our normal understanding of true sentences, they do not name the extension of the concept named by the horizontal, just as it is immaterial that, on our normal understanding of numbers, they do not name extensions of concepts. The True is what is common to all true thoughts, just as a number is what is common to all concepts that are equinumerate. The point is that the assignment of a particular extension as what is named by a true sentence (numeral) is adequate for purposes of science—that is, it both satisfies all the laws of truth (number) and determines a Bedeutung for any complex expression constructed out of the sentence (numeral) and other terms that have Bedeutung.

Frege assumes his readers have the ability to recognize the logical composition of everyday thoughts that they already grasp. And it is this assumption that licenses his belief that they will understand his introductions of primitive function-names in *Basic Laws*. Although Frege's decomposition is somewhat different from those his readers antecedently recognized, his elucidations are supposed to show that he is truly doing nothing more than introducing new, more precise, decompositions and that the constituents, on his decomposition, really have the precision required by logic, that is, really have Bedeutung. The constraints on expression imposed by Frege's logical notation require that sentential expressions, expressions that express thoughts, have the same logical status as proper names that do not express thoughts. This requires the fixing of the sense of identity statements in which the proper name on one side does not express a thought and the proper name on the other side does express a thought. Section 10 is supposed to give us this additional information about the Bedeutung of sentential expressions.

An upshot is that every Begriffsschrift expression (i.e., every precise expression) for the True is complex. One might object that, since Frege says truth is primitive and simple and that truth is indefinable (PW 126–129, PW 174, CP 353), it is illegitimate for him to use complex symbols for it. But it is not clear that Frege has done this. Frege does not identify *the True* as something simple, primitive, and indefinable but, rather, *the predication of truth*. A name for this simple logical notion would not be a proper name but a function-name. It is less clear that the True is logically simple. A direction is what is common to all lines parallel to a particular line, and a particular extension is what is common to all concepts that hold of precisely those objects of which some particular concept holds. In the same way, the True is what is common to all thoughts that are true.

Furthermore, in most of Frege's discussions, logical simplicity is regarded as a feature not of concepts and objects but, rather, of senses. He seems to suggest that what is logically simple cannot be analyzed (see, e.g., CP 302, PW 208), and that the analysis of what is complex is used to formulate definitions. But this analysis is analysis of senses. And the senses that pick out truth-values, since they are thoughts, are obviously complex. It is not entirely clear how one would characterize the logical simplicity of an object, but an obvious possibility is that it could not be constructed out of other entities. Such a view is suggested by Frege's metaphorical remark in "On Sense and Meaning," "One might also say that judgements are distinctions of parts within truth-values" (CP 165). Although I would not be inclined to draw any serious consequences from this remark, it does suggest that, if logical simplicity is a feature of parts of the realm of reference, the True is not logically simple.[2]

It should be clear, at this point, that there is no hidden primitive proper name in Frege's logic. It should also be clear that Frege is not giving a circular or defective argument that all his simple names have Bedeutung. Rather, he is simply giving hints about the possible decomposition of thoughts which are designed to get his readers to see how his simple names are to be understood. From such an understanding of his simple names, it should be clear that they do have Bedeutung. These hints, of course, depend crucially on the legitimacy of his primitive logical laws, including Basic Law V. Thus, it is not surprising that he says, in his response to the letter in which Russell introduces the contradiction that follows from Basic Law V,

> It seems accordingly that the transformation of the generality of an identity into an identity of ranges of values (sect. 9 of my *Basic Laws*) is not always permissible, that my law V (sect. 20, p. 36) is false, and that *my explanations in sect. 31 do not suffice to secure a meaning for my combination of signs in all cases.* (COR 132, my emphasis)

[2]As I have argued, in chapter 5, it is false that Frege thought there were logically simple logical objects, and it is unlikely even that he believed that there were any logically simple objects.

Absolute Simples

Although I have raised the issue of how to understand Frege's notion of primitiveness at a number of points, I have not at any point tried to give an account of Fregean primitiveness. This is partly because Frege says so little about primitiveness. In spite of the rather crucial role primitiveness plays, Frege does not devote the kind of effort to explaining it that he devotes to such notions as concepthood or truth. One of the reasons may be that, although his readers would not be in a position to acknowledge the success of his project without an understanding of his notion of concepthood, a similar understanding of primitiveness is not required. As long as his readers feel that they understand the primitive terms Frege introduces in his Begriffsschrift and agree that these primitive terms express basic logical notions (and as long as the inconsistency of Basic Law V is unnoticed), they can acknowledge the success of Frege's introduction of a purely logical science that can replace arithmetic. I want to look more closely at Fregean primitiveness here.

As Frege's discussions of elucidation suggest, it would be a mistake to take primitiveness as some inherent property of linguistic symbols. From the point of view of someone using a particular logically perfect language, certain symbols are primitive terms. But from the point of view of someone constructing a logically perfect language for a systematic science, there is no requirement that particular symbols be used as the primitive terms. The primitiveness of a term must have something to do with its sense or Bedeutung. It seems likely that primitiveness has something to do with the subject matter of the systematic science under construction.

It is also important to remember that a systematic science will not be constructed arbitrarily. Frege introduces the notion of systematic science as a refinement of what demands should be placed on the sciences. His aim is not to get his readers to invent new sciences but, rather, to convince them that there is something to be gained by introducing a systematic science that can replace the everyday presystematic sciences (or at least everyday presystematic mathematics). In order to do this, investigators must look at the terms of the presystematic science and explicitly define those that are complex. Presumably, at some point in the process of attempting to define a complex term, they will reach a point at which there is nothing more to say, at which the senses of the terms reached seem too simple to require explanation. These will be the primitive terms.

What constraints are there on such a procedure? At the very least, what must be preserved, in the new systematic science, are the basic inferences of the old, presystematic science. When Frege undertakes to provide a logically perfect language for arithmetic, for instance, this is one of the tests of his definitions. When the arithmetical symbols that appear in a sentence that was regarded as expressing a basic truth of presystematic arithmetic are replaced by their Fregean definitions, the result must express something that is provable in Frege's systematic science of arithmetic. Similarly, basic inferences must be preserved. From the thought expressed by '2 is even' (when all constituents of the sentence are taken as abbreviations of Fregean definitions), Frege's systematic science must licence us to infer the thought expressed by 'Every multiple of 2 is even' (when all all constituents are understood in this way).

The constraint described above prevents us from taking some scientific terms as primitive. If, for instance, from the claim that all horses are mammals we are entitled immediately to infer that all horses are vertebrates, then the concepts of mammal and vertebrate cannot both be simple elements of the science. On the other hand, it is not obvious that the constraints of preserving basic inferences and truths of science dictate precisely which are the simple elements of science. Does Frege hold such a view?

It is tempting to suppose that Frege's search for the primitive terms of arithmetic and his claim that it takes scientific work to reach what is logically simple (CP 182) indicate a belief that there is a set of absolute simples that underlie all scientific thought or comprise the ultimate constituents of reality.[1] If so, then it may be that we can expect an examination of the logical structure of our sciences to yield these simples. And even if such an examination is not itself sufficient, it may be that our understanding of the most general science (logic) will enable us to distinguish between logical simples and complex elements. In either case a doctrine of absolute simples can explain the success of elucidation. For if an understanding of these absolute simples underlies everyone's reasoning, it must be possible to come to a common understanding of terms for these simples. Given this, the necessity of basing this mutual understanding on guessing and figurative modes of expression will not undermine Frege's requirements of precision. For these primitive terms will merely be terms for what is basic to all our communication.

Some of the strongest support for the view of primitive terms as names for absolute simples comes from passages from "On Concept and Object" where Frege talks about the impossibility of decomposing what is logically simple. He says this as part of an explanation of his refusal to define the notion of concept. It seems likely that he thinks that an understanding of the logical role played by concepts underlies our common understanding of correct inference and, since concepts are simply elements that play this logical role, everyone already understands what concepts are. It remains only to coin a term for this notion and to make sure that his audience knows what he is talking about. But even if this is an accurate description of Frege's discussion of his notion of concept, and even if we infer that Frege is regarding the notion of concepthood as a logical simple, this is not sufficient to show that a set of terms for all logical simples will suffice for the construction of all systematic sciences; nor is it sufficient to show that all primitive terms must be words for absolute logical simples.

Additional support for attributing a doctrine of absolute simples appears in the following passage.

> For not everything can be defined; only what has been analysed into concepts can be reconstituted out of the parts yielded by the analysis. But what is simple cannot be analysed and hence not defined. (PW 271)

---

[1]This view is defended by E.-H. Kluge in *The Metaphysics of Gottlob Frege*, 101–108. Kluge also argues that there are no logically simple objects and, hence, no primitive proper names.

And, in a pre-*Foundations* discussion of Boole's logical notation, he says, in a discussion of the logical structure of the claim that $2^x = 16$,

> Of course, if the expression of the content of possible judgement is to be analysable in this way, it must already be itself articulated. We may infer from this that at least the properties and relations which are not further analysable must have their own simple designations. (PW 17)

The picture that may seem implicit in these passages is one on which, by analyzing the thoughts (the senses of the sentences used) which constitute our everyday science, we will ultimately reach the simple constituents of these thoughts. These simple constituents of the thoughts pick out (or are modes of presentation of) simple elements. If primitive terms are terms that stand for these simple elements, then Frege seems committed to the view that there is a determinate group of primitive terms with logically simple Bedeutung from which all terms of legitimate science can be constructed.

Frege clearly believes that there are some concepts that are logically simple concepts in an absolute sense, that is, that underlie all legitimate science—the concept of function is an example—but there is also evidence that he did not hold the doctrine of simples described above. That is, there is evidence that the construction of systematic sciences requires the use of some primitive terms that stand for simples that are not logically simple in an absolute sense but are only logically simple with respect to a particular systematic science. It will be useful to examine this evidence in detail, including some apparent evidence that does not hold up on close inspection.

Some of the less conclusive evidence comes from two passages in which Frege talks about alternative analyses of thoughts.[2] One of these is from his correspondence. In a letter dated August 29, 1882, Frege writes, "I do not believe that for any judgeable content there is only one way in which it can be decomposed, or that one of these possible ways can always claim objective pre-eminence" (COR 101). One might suspect that on the doctrine of absolute simples there could be only one correct logical decomposition and that this decomposition would be exhibited by a unique expression of the judgeable content in Begriffs-

---

[2]The following evidence—both of the passages in which Frege talks about alternative analyses of thoughts and about alternative systems—is discussed by Hans Sluga in "Frege against the Booleans." Sluga appears to think the tensions between these passages in Frege's talk about simples cannot be resolved.

schrift notation using only primitive terms. But Frege does not deny this in the letter quoted. His next sentence is: "In the inequality 3 > 2 we can regard either 2 or 3 as the subject." The decomposition into subject and predicate is not a logical decomposition. Thus, it does not follow from Frege's remark that there are distinct Begriffsschrift expressions of this thought—for a Begriffsschrift formula has no subject. Frege's point here is that once one truly understands logical structure—that is, sees that Begriffsschrift notation captures all logical structure—apparently different analyses of a particular thought can be seen not to be distinct for scientific (logical) purposes. Not only are the primitive terms involved in the two analyses identical, the entire Begriffsschrift structure is identical. This seems also to fit with his discussions of apparently different Begriffsschrift expressions of certain thoughts.

In the discussion that continues in the same letter, Frege shows how existential judgements are to be represented in Begriffsschrift notation. There is, in this discussion, one example of different Begriffsschrift formulas that express the same thought. Frege notes that one can add, to a formula, two negation-strokes that "cancel each other" without changing the thought expressed. It is not clear that this example tells against the doctrine of absolute simples, for the different Begriffsschrift expressions of a thought are equivalent, and at least the nonlogical signs for simples will be the same in each expression. Furthermore, this does not show that one of Frege's primitive logical symbols, the negation-stroke, can be eliminated in all Begriffsschrift expressions; it only allows the elimination of pairs of negation-strokes. Thus, there is no reason to suppose that this observation has anything to do with the existence of alternate groups of underlying logical simples.

The second passage in which Frege discusses alternative analyses is similarly inconclusive. In "On Concept and Object" Frege claims that the sentences

(a) there is at least one square root of 4

and

(b) the concept square root of 4 is realized

express the same thought, but that (a) is about a concept while (b) is about an object. One might suppose, then, that the Begriffsschrift expressions of (a) and (b) will be different and not equivalent. In fact, if the expression 'the concept square root of 4' stands for the extension of the concept predicated in (a), these two expressions of the thought will be provably equivalent in Frege's later version of Begriffsschrift.[3] Furthermore, with one exception, the same primitive terms are involved in the distinct expressions of the thought. The exception is the notation that allows Frege to turn concept-expressions into object-expressions that stand for the extension of the concept. This is not evidence of the existence of alternative groups of simples. For all the primitive terms that appear in either expression must be used in an adequate logical notation. The demonstration of this equivalence may simply be viewed as elucidating the content of the primitive terms.

It seems entirely plausible that when Frege says that there are alternative analyses of a thought, the difference between the alternatives is either psychological or presystematic, not part of a systematic science. From the point of view of the systematic science, some of these analyses (those, for instance, which differ in having different subjects) cannot be distinguished at all. Others can be distinguished only by their typographically distinct (although logically equivalent) Begriffsschrift expressions. It is compatible with Frege's discussions of alternative analyses that there be a Begriffsschrift formula that expresses the thought and to which every other Begriffsschrift expression is provably equivalent. Furthermore, although each of the different analyses in Frege's examples may be expressible using a different set of primitive terms, neither set is sufficient on its own for expressing all logical truths. Thus, this also seems compatible with the view that the primitive symbols of Begriffsschrift correspond to the unique set of logically simple elements of the most general science.

There are, however, more convincing reasons to avoid attributing the doctrine of simples to Frege. In "Logic in Mathematics" he says:

> We can see from this that the possibility of one system does not necessarily rule out the possibility of an alternative system, and that we may

---

[3]The two expressions will be

$$\neg\cup\neg\, a^2 = 4$$

and

$$\neg\cup\neg \setminus \grave{\alpha} \left(\neg\!\mathscr{S}\!\top\, g(\alpha) = \alpha \atop \grave{\varepsilon}\,(\varepsilon^2 = 4) = \grave{\varepsilon}\, g(\varepsilon) \right)$$

These are equivalent, but the Begriffsschrift proof that they are equivalent, as section 52 of *Basic Laws* shows, requires Basic Law V. It is unlikely that Frege would have made the same remark after the collapse of his system.

have a choice between different systems. So it is really only relative to a particular system that one can speak of something as an axiom. (PW 206)

And in the introduction to *Basic Laws*:

Anyone who holds other convictions has only to try to erect a similar structure upon them, and I think he will perceive that it does not work, or at least *does not work so well*. (BLA 25, my emphasis)

It seems plausible that Frege is leaving open the possibility that there are alternative systematic constructions of arithmetic. This suggests that the notion of primitive term is relative to a particular system.

This suggestion is more strongly supported by some of Frege's discussions of the primitive terms of logic. For Frege did recognize the existence of alternative sets of primitive logical terms. In an unpublished article, "Boole's Logical Calculus and the Concept-Script," Frege discusses his choice of the conditional-stroke as a primitive logical term. He says that the use of the conditional-stroke between *A* and *B* amounts to the denial of the third member of the list:[4]

  I *A* and *B*
  II *A* and not *B*
  III not *A* and *B*
  IV not *A* and not *B* (PW 35)

He also says:

Now to obtain a sign joining two contents of possible judgement whose meaning was as simple as possible, I had four choices open, all from this point of view equally justified: I could have adopted as the meaning of such a sign the denial of any one of the four cases mentioned above. But it sufficed to choose one, since the four cases can be converted into one another by replacing A and B by their denials. To use a chemical metaphor, they are only allotropes of the same element. I chose the denial of the third case, because of the ease with which it can be used in inference, and because its content has a close affinity with the important relation of ground and consequent. (PW 37)

Frege recognizes that he could have introduced a primitive sign for the denial of the fourth case, that is, disjunction. But a symbol for disjunc-

---

[4]Frege's conditional-stroke is written in such a way that the consequent of the conditional appears above the antecedent. Thus, if we assign letters in alphabetical order from top to bottom, 'A' will be assigned to the consequent and 'B' to the antecedent.

tion can be defined from the negation-stroke and the conditional-stroke. Thus, if the notion of logical simplicity is absolute, disjunction cannot be logically simple. Nonetheless, Frege's objections to a logical notation that has no conditional-stroke and has a primitive sign for disjunction have to do, not with claims about the logical complexity of disjunction, but with such psychological concerns as ease of use and affinity with an important relation. Frege repeats this view in one of his last publications, "Compound Thoughts." He says there his choice "is not governed by any fact of logic" (CP 404).

This fits with the description of Frege's project I have given. For the definitions of the symbols of arithmetic are constrained only by Frege's desire to use them to construct a systematic arithmetic that has all the applications of the traditional nonsystematic arithmetic. But Frege does not seem to think that this constraint dictates a set of available simples from which the objects of arithmetic must be constructed. One of the simple elements in his definitions of the numbers is the notion of the extension of a concept. But he says of these definitions, in *Foundations*,

> In this definition the sense of the expression "extension of a concept" is assumed to be known. This way of getting over the difficulty cannot be expected to meet with universal approval, and many will prefer other methods of removing the doubt in question. I attach no decisive importance even to bringing in the extensions of concepts at all. (FA 117)

Although it is true that, after learning of Russell's paradox and realizing the difficulties inherent in his notion of extension, he despairs of defining the numbers, his reason is not that the notion of extension is a simple element underlying all thought.

Frege's general notion of primitiveness is a notion of primitiveness-relative-to-a-system. There is no assumption that there is a unique set of simple elements of reality which must be used in the construction of any systematic science. It does not follow, however, that *anything* can be introduced as a primitive element of some science. As I mentioned earlier, the logical structure of a science rules out certain terms as primitive. Furthermore, there are other constraints on the introduction of primitive terms. For instance, it must be possible, by means of elucidation, to eliminate any disagreement about the sense of the term and to indicate the ultimate grounds that will be necessary for the justification of thoughts in whose expression the term appears. This, of course, constitutes one respect in which primitive terms have logically simple content. There is also another respect. From the point of view of the systematic science in whose expressions they appear as primitive terms, their content is logically simple. For example, the conditonal-stroke is a primitive term in Frege's Begriffsschrift and, since the other truth-functional connectives can be defined by use of primitive Begriffsschrift signs (the conditional-stroke and negation-stroke), it is not permissible to introduce a primitive Begriffsschrift symbol for any other truth-functional connective. As a consequence, it is not possible to construct a symbol that can replace the conditional-stroke using only the other symbols of Begriffsschrift.

It does not follow from the existence of alternate sets of primitive elements which can be used to construct some systematic science that there are no absolute simples at all. It follows only that a set of terms for the absolute simples will not be sufficient for the construction of all language needed to express legitimate systematic science. All terms for absolute simples must, of course, be primitive, but not all primitive terms must be terms for absolute simples.[5] Except for the few cases in which the primitive term is to stand for an absolute simple, the efficacy of elucidation cannot be guaranteed by our common understanding of correct inference. This, of course, is needed because such an understanding, along with a recognition of the basic truths and inferences of some science, will be needed if we are to come to an agreement about the basic elements of the science. But something else is required, as well. We must also have a common understanding of the subject matter of the science. The upshot is that Fregean elucidation really is as imprecise as it seems in the passages with which I began.

---

[5]Kluge disagrees. It seems plausible that Kluge's reason for denying the possibility that primitiveness is relative to a system is his emphasis on Frege's ontology rather than on Frege's epistemology. On Kluge's view of Frege's project, the construction of a theory of metaphysics is central. As a result, the significance of primitiveness cannot simply be what is required by Frege's epistemological constraints, and the only obvious evidence for interpreting primitiveness is the evidence provided by Frege's passages concerning logical simplicity.

Bibliography of Works Cited

Frege's Writings

The Basic Laws of Arithmetic: Exposition of the System. Translated and edited with an introduction by Montgomery Furth. Berkeley and Los Angeles: University of California Press, 1964.

Begriffsschrift, a Formula Language, Modeled upon that of Arithmetic, for Pure Thought. In *Frege and Godel: Two Fundamental Texts in Mathematical Logic.* Edited by Jean van Heijenoort. Cambridge, Mass.: Harvard University Press, 1970.

Begriffsschrift und andere Aufsätze. Edited by Ignacio Angelelli. Hildesheim: Olms, 1964. [BS]

Collected Papers on Mathematics, Logic, and Philosophy. Edited by Brian McGuinness. Translated by Max Black, V. H. Dudman, Peter Geach, Hans Kaal, E.-H. W. Kluge, Brian McGuinness, R. H. Stoothoff. Oxford: Blackwell, 1984.

Conceptual Notation and Related Articles. Translated and Edited by Terrell Ward Bynum. Oxford: Clarendon, 1972.

The Foundations of Arithmetic. Translated by J. L. Austin. 2d rev. ed. Evanston, Ill.: Northwestern University Press, 1980.

Grundgesetze der Arithmetik. Hildesheim: Olms, 1962.

Kleine Schriften. Edited by Ignacio Angelelli. Hildesheim: Olms, 1967.

Nachgelassene Schriften. Edited by Hans Hermes, Friedrich Kambartel, and Friedrich Kaulbach. Hamburg: Meiner, 1969.

Philosophical and Mathematical Correspondence. Edited by Gottfried Gabriel, Hans Hermes, Friedrich Kambartel, Christian Thiel, Albert Veraart. Abridged from the German edition by Brian McGuinness. Translated by Hans Kaal. Chicago: University of Chicago Press, 1980.

Posthumous Writings. Edited by Hans Hermes, Friedrich Kambartel, Friedrich Kaulbach. Translated by Peter Long and Roger White. Chicago: University of Chicago Press, 1979.

Translations from the Philosophical Writings of Gottlob Frege. Edited by Peter Geach and Max Black. Index by E. D. Klemke. 3d ed. Totowa, N.J.: Rowman and Littlefield, 1980.

Wissenschaftlicher Briefwechsel. Edited by Gottfried Gabriel, Hans Hermes, Friedrich Kambartel, Christian Thiel, and Albert Veraart. Hamburg: Meiner, 1976.

Other Writings

Baker, G. P., and P. M. S. Hacker. *Frege: Logical Excavations.* New York: Oxford University Press, 1984.

Bell, David. *Frege's Theory of Judgement.* Oxford: Clarendon, 1979.

Benacerraf, Paul. "Frege: The Last Logicist." In *Midwest Studies in Philosophy*, vol. 6. Edited by Peter A. French, Theodore E. Uehling, Jr., and Howard K. Wettstein. Minneapolis: University of Minnesota Press, 1981.

——. "Mathematical Truth." In *Philosophy of Mathematics: Selected Readings.* Edited by Paul Benacerraf and Hilary Putnam. 2d ed. Cambridge: Cambridge University Press, 1983.

Benacerraf, Paul, and Hilary Putnam, eds. *Philosophy of Mathematics: Selected Readings.* Cambridge: Cambridge University Press, 1983.

Bostock, David. *Logic and Arithmetic: Natural Numbers.* Oxford: Clarendon, 1974.

——. *Logic and Arithmetic: Rational and Irrational Numbers.* Oxford: Oxford University Press, 1979.

Burge, Tyler. "Frege on Extensions of Concepts from 1884 to 1903." *Philosophical Review*, 93 (1984), 3–34.

——. "Frege on Truth." In *Frege Synthesized.* Edited by Leila Haaparanta and Jaakko Hintikka. Dordrecht: Reidel, 1986.

——. "Sinning against Frege." *Philosophical Review*, 88 (1979), 398–432.

Bynum, Terrell Ward, trans. and ed. *Gottlob Frege: Conceptual Notation and Related Articles.* Oxford: Clarendon, 1972.

Carnap, Rudolf. *The Logical Foundations of Probability.* Chicago: University of Chicago Press, 1962.

Currie, Gregory. *Frege: An Introduction to His Philosophy.* Totowa, N.J.: Barnes and Noble, 1982.

Dedekind, Richard. *Essays on Number.* Translated by Wooster Woodruff Beman. Chicago: Open Court, 1901.

Diamond, Cora. "Frege against Fuzz." In a collection of her papers forthcoming from MIT Press.

——. "Throwing Away the Ladder." *Philosophy*, 63 (1988), 5–27.

——. "What Does a Concept-Script Do?" In *Frege: Tradition and Influence.* Edited by Crispin Wright. Oxford: Blackwell, 1984.

Dudman, V. H. "The Concept Horse." *Australasian Journal of Philosophy*, 50 (1972), 67–76.

Dummett, Michael. *Frege: Philosophy of Language.* New York: Harper and Row, 1973.

——. *The Interpretation of Frege's Philosophy.* Cambridge, Mass.: Harvard University Press, 1981.

——. "An Unsuccessful Dig." In *Frege: Tradition and Influence.* Edited by Crispin Wright. Oxford: Blackwell, 1984.

Geach, Peter. "Frege." In *Three Philosophers*. G. E. M. Anscombe and Peter Geach. Oxford: Blackwell, 1961.

——. "Saying and Showing in Frege and Wittgenstein." In *Essays in Honour of G. H. von Wright*. Edited by Jaakko Hintikka. *Acta Philosophica Fennica*, 28. Amsterdam: North-Holland, 1976.

Gillies, Donald. *Frege, Dedekind, and Peano on the Foundations of Arithmetic*. The Netherlands: Van Gorcum, 1982.

Gödel, Kurt. "What Is Cantor's Continuum Problem?" In *Philosophy of Mathematics: Selected Readings*. Edited by Paul Benacerraf and Hilary Putnam. Cambridge: Cambridge University Press, 1983.

Haaparanta, Leila. *Frege's Doctrine of Being*. Acta Philosophica Fennica, 39. Helsinki: Philosophical Society of Finland, 1985.

Haaparanta, Leila, and Jaakko Hintikka, eds. *Frege Synthesized*. Synthese Library, vol. 181. Dordrecht: Reidel, 1986.

Heijenoort, Jean van. "Frege and Vagueness." In *Frege Synthesized*. Edited by Leila Haaparanta and Jaakko Hintikka. Dordrecht: Reidel, 1986.

——. "Logic as Calculus and Logic as Language." *Synthese*, 17 (1967).

Hodes, Harold. "Logicism and the Ontological Commitments of Arithmetic." *Journal of Philosophy*, 81 (1984), 123–149.

Hugly, Philip. "The Ineffability in Frege's Logic." *Philosophical Studies*, 24 (1973), 227–244.

Jeffrey, Richard. "Dracula Meets Wolfman: Acceptance vs. Partial Belief." In *Induction, Acceptance, and Rational Belief*. Edited by M. Swain. Dordrecht: Reidel, 1970.

——. "Probable Knowledge." In *The Problem of Inductive Logic*. Edited by I. Lakatos. Amsterdam: North-Holland, 1968.

——. "Valuation and the Acceptance of Scientific Hypotheses." *Philosophy of Science*, 23 (1956).

Kahneman, Daniel, Paul Slovic, and Amos Tversky, eds. *Judgment under Uncertainty: Heuristics and Biases*. Cambridge: Cambridge University Press, 1982.

Kant, Immanuel. *Critique of Pure Reason*. Translated by Norman Kemp Smith. London: Macmillan, 1929.

——. *Logic*. Translated by Robert S. Hartman and Wolfgang Schwarz. New York: Bobbs-Merrill, 1974.

Kitcher, Philip. "Frege, Dedekind, and the Philosophy of Mathematics." In *Frege Synthesized*. Edited by Leila Haaparanta and Jaakko Hintikka. Dordrecht: Reidel, 1986.

——. "Frege's Epistemology." *Philosophical Review*, 88 (1979), 235–262.

——. *The Nature of Mathematical Knowledge*. New York: Oxford University Press, 1983.

Klemke, E. D., ed. *Essays on Frege*. Chicago: University of Illinois Press, 1968.

Kline, Morris. *Mathematical Thought from Ancient to Modern Times*. New York: Oxford University Press, 1972.

Kluge, E.-H. W. *The Metaphysics of Gottlob Frege*. The Hague: Martinus Nijhoff, 1980.

Kripke, Saul. "Naming and Necessity." In *Semantics of Natural Language*. Edited by Donald Davidson and Gilbert Harman. Dordrecht: Reidel, 1972.

Maddy, Penelope. "Perception and Mathematical Intuition." *Philosophical Review*, 89 (1980), 163–196.

Moore, A. W., and Andrew Rein. "*Grundgesetze*, Section 10." In *Frege Synthesized*. Edited by Leila Haaparanta and Jaakko Hintikka. Dordrecht: Reidel, 1986.

Parsons, Charles. "Frege's Theory of Number." In Parsons, *Mathematics and Philosophy*. Ithaca, N.Y.: Cornell University Press, 1983.

Parsons, Terence. "Why Frege Should Not Have Said 'The Concept *horse* is not a Concept.'" *History of Philosophy Quarterly*, 3 (1986), 449–465.

Resnik, Michael D. *Frege and the Philosophy of Mathematics*. Ithaca, N.Y.: Cornell University Press, 1980.

Ricketts, Thomas. "Generality, Meaning, and Sense." *Pacific Philosophical Quarterly*, 67 (1986), 172–195.

——. "Objectivity and Objecthood: Frege's Metaphysics of Judgment." In *Frege Synthesized*. Edited by Leila Haaparanta and Jaakko Hintikka. Dordrecht: Reidel, 1986.

Russell, Bertrand. *Introduction to Mathematical Philosophy*. New York: Simon and Schuster, 1960.

Schröder, E. Review of Frege's *Conceptual Notation*. *Zeitschrift für Mathematik und Physik*, 25 (1880). Translated and reprinted in *Gottlob Frege: Conceptual Notation and Related Articles*. Translated and edited by Terrell Ward Bynum. Oxford: Clarendon, 1972.

Sluga, Hans. "Frege against the Booleans." *Notre Dame Journal of Formal Logic*, 28 (1987).

——. *Gottlob Frege*. London: Routledge and Kegan Paul, 1980.

Steiner, Mark. *Mathematical Knowledge*. Ithaca, N.Y.: Cornell University Press, 1975.

Venn, J. Review of Frege's *Conceptual Notation*. *Mind*, 5 (1880). Reprinted in *Gottlob Frege: Conceptual Notation and Related Articles*. Translated and edited by Terrell Ward Bynum. Oxford: Clarendon, 1972.

Weiner, Joan. "The Philosopher behind the Last Logicist." In *Frege: Tradition and Influence*. Oxford: Blackwell, 1984.

Wright, Crispin. *Frege's Conception of Numbers as Objects*. Aberdeen, Scotland: Aberdeen University Press, 1983.

Wright, Crispin, ed. *Frege: Tradition and Influence*. Oxford: Blackwell, 1984.

Index

Library of Congress Cataloging-in-Publication Data

Weiner, Joan.
    Frege in perspective / Joan Weiner.
      p.   cm.
    Includes bibliographical references.
    ISBN 0-8014-2115-2 (alk. paper)
    1. Frege, Gottlob, 1848–1925.   I. Title.
  B3245.F24W45   1990
  193—dc20   89-28377